Lecture Notes in Economics and Mathematical Systems

571

T0223702

Tobias Herwig

Market-Conform
Valuation of Options

 Springer

Author

Tobias Herwig
Graduate Program "Finance and Monetary Economics"
Faculty of Economics and Business Administration
Johann Wolfgang Goethe University
Mertonstrasse 17–21
60054 Frankfurt am Main
Germany
TobiasHerwig@web.de

ISSN 0075-8442
ISBN-10 3-540-30837-7 Springer Berlin Heidelberg New York
ISBN-13 978-3-540-30837-9 Springer Berlin Heidelberg New York

Springer is a part of Springer Science+Business Media

springer.com

© Springer-Verlag Berlin Heidelberg 2006
Printed in Germany

Typesetting: Camera ready by author
Cover design: *Erich Kirchner*, Heidelberg

Printed on acid-free paper   42/3153DK   5 4 3 2 1 0

Acknowledgements

The present study emanated from my participation in the graduate program 'Finance and Monetary Economics' at the Johann Wolfgang Goethe–University Frankfurt am Main, Germany. It was accepted as a doctoral thesis by the Faculty of Economics and Business Administration in June 2005.

It is a great pleasure to thank all the people who helped and supported me during the last years. First and foremost I would like to thank my Ph.D. supervisor Prof. Christian Schlag who introduced me to the field of derivatives and financial engineering, who gave me valuable advice and support and who created the ideal environment for productive research at the graduate program in Frankfurt.

I would also like to thank Prof. Raimond Maurer for acting as the second thesis supervisor and committee members Prof. Uwe Hassler and Prof. Dieter Nautz for reviewing my thesis and for their interest in my work.

I am very grateful to my colleagues from the graduate program and the Chair of Derivatives and Financial Engineering, Michael Belledin, Christoph Benkert, Carsten Bienz, Silke Brandts, Nicole Branger, Angelika Esser, Iskra Kalodera, Micong Klimes, Keith Küster, Burkart Mönch, and Alexander Schulz for many fruitful discussions and their valuable feedback.

My special thanks go to Christian Offermanns and Karsten Ruth, who gave me the opportunity to discuss ideas with them and who helped to create an enjoyable working atmosphere in our office. I am deeply grateful to them for investing time and energy proofreading multiple versions of my work and providing helpful comments.

VI   Acknowledgements

For financial support, I would like to thank the Deutsche Forschungs-gemeinschaft (DGF), who supported me through the graduate program and the DZ Bank Stiftung who funded the publication of this monograph.

Frankfurt am Main, December 2005   *Tobias Herwig*

Contents

1

Introduction

1.1 The Area of Research

In this thesis, we will investigate the 'market-conform' pricing of newly issued contingent claims. A contingent claim is a derivative whose value at any settlement date is determined by the value of one or more other underlying assets, e.g., forwards, futures, plain-vanilla or exotic options with European or American-style exercise features. Market-conform pricing means that prices of existing actively traded securities are taken as given, and then the set of equivalent martingale measures that are consistent with the initial prices of the traded securities is derived using no-arbitrage arguments. Sometimes in the literature other expressions are used for 'market-conform' valuation – 'smile-consistent' valuation or 'fair-market' valuation – that describe the same basic idea.

The seminal work by Black and Scholes (1973) (BS) and Merton (1973) mark a breakthrough in the problem of hedging and pricing contingent claims based on no-arbitrage arguments. Harrison and Kreps (1979) provide a firm mathematical foundation for the Black–Scholes–Merton analysis. They show that the absence of arbitrage is equivalent to the existence of an equivalent martingale measure. Under this measure the normalized security price process forms a martingale and so securities can be valued by taking expectations. If the securities market is complete, then the equivalent martingale measure and hence the price of any security are unique. If the market is not complete, a much more realistic assumption in practice, this will no longer hold, so that the investor has to decide how to pick the equivalent martingale measure to be used for pricing.

The approaches in the literature can be divided into two main classes. The first class starts with an assumption about the data-generating process, i.e. about the stochastic process that drives the underlying asset price. The most popular choice for the data-generating process is a geometric Brownian motion, first applied in option pricing theory by Black and Scholes (1973). However, the behavior of implied volatilities derived from inverting the Black–Scholes formula, makes the validity of this model questionable. The empirical evidence provided by, among others, Rubinstein (1994), Jackwerth and Rubinstein (1996), Dumas et al. (1998), or Aït-Sahalia and Lo (1998) shows that implied volatilities vary across different strikes (i.e. they exhibit a smiles or skews pattern) and different times to maturity (term structure), while the BS model does not allow for such variations. These variations can roughly be explained by more sophisticated models, such as stochastic volatility (e.g. Hull and White (1987), Heston (1993), Schöbel and Zhu (1999)), stochastic interest rates (e.g. Merton (1973), Amin and Jarrow (1992)), jump models (e.g. Merton (1976), Bates (1991)), or combinations of the different processes (e.g. Bates (1996), Scott (1997), Bakshi and Chen (1997)). After defining a stochastic process for the underlying, this process has to be rewritten in risk-neutral terms. Then, the parameters of the processes for the underlying asset price and for the volatility and/or jump process are estimated. Most calibration procedures rely on the existence of explicit pricing formulas for the prices of benchmark instruments, since the unknown parameters are found by inverting such pricing formulas. When closed-form expressions exist, the model parameters can often be simply estimated by employing least-squares methods. However, closed-form solutions for prices are not always available or easy-to-compute. In this case, fitting the model to market prices implies searching the parameter space via direct simulation, which is computationally expensive and time-consuming. Finally, after specifying the model parameters of the stochastic processes, the prices of new contingent claims are derived as a function of the parameters of these processes and the price of the underlying asset.

Unfortunately, these models often do not fit observed market prices accurately (e.g. Das and Sundaram (1999), Belledin and Schlag (1999)). Therefore, they should be used carefully in practice, especially to price and hedge exotic options. This is due to the fact that in order to improve the hedging performance, exotic and standard options need to be valued consistently, since exotic options are often hedged with portfolios of European options. These problems are discussed in the literature on 'market-conform' or 'smile-consistent' no-arbitrage models, the second class of no-arbitrage approaches.

Market-conform models reverse the approach followed in the conventional stochastic volatility or jump models. The prices of actively traded European options are taken as given, and they are used to infer information about the underlying price process. The implementation of market-conform models for pricing and hedging purposes is mainly done in a discrete time framework. The tools used are either implied binomial/trinomial trees, implicit finite difference schemes, or weighted Monte Carlo simulations.

The most popular 'market-conform' approaches are the so-called implied tree models, which are extending the seminal binomial model of Cox et al. (1979). In the standard Cox et al. (1979) tree, the size of the up and down move of the underlying and the respective transition probability of such moves are constant, since they depend on the volatility, which is assumed to be constant. This is no longer the case for implied trees. Implied binomial (or trinomial) trees are built from the known prices of European options. In order to build a consistent risk-neutral price process of the underlying, these exchange-traded options are used to infer information about the data-generating process. They are called 'implied trees', because they are consistent with or implied by the volatility structure and can be viewed as a discretization of generalized one-dimensional diffusions in which the volatility parameter is allowed to be a function of both time and asset price.

We propose a new method to construct arbitrage-free implied binomial trees based on the approach by Brown and Toft (1999). As the output of our procedure we get an arbitrage-free, risk-neutral implied binomial tree, which is consistent with the term structure of implied volatilities and also with the implied volatility smile. The implied risk-neutral probability distributions (IRNPDs) for later maturity dates are an endogenous result of the model and take the IRNPDs of the prior maturity dates into account. Our method can also be used to construct arbitrage-free, risk-neutral implied multinomial trees. This multinomial setting can be used to calibrate models with more than one state variable, e.g. the underlying price process and stochastic volatility. Since the approaches suggested by Rubinstein (1994), Dupire (1994), Derman and Kani (1994), Derman et al. (1996), Jackwerth (1997), Barle and Cakici (1998), and Brown and Toft (1999) are closely related to our new technique, we briefly describe the differences and, in particular, the drawbacks of these models. For more detailed surveys see Jackwerth (1999) or Skiadopoulos (2001).

The key idea of the approach suggested by Rubinstein (1994) is the estimation of the IRNPD at the terminal date of the tree. This IRNPD is close to a prior guess subject to some constraints. There are two major drawbacks of this method. First, the implied binomial tree fits the strike dimension of the volatility smile only at one single maturity date and neglects information for traded options with shorter or longer maturities. This leads directly to the second problem, namely the fact that implied binomial trees constructed for two different maturity dates are not necessarily consistent for overlapping time-periods. To overcome these problems, Jackwerth (1997) develops a generalized implied binomial tree by introducing a piecewise-linear weight function and by using nodal probabilities instead of path probabilities. This generalization allows for the incorporation of all the information and fits the complete volatility surface. However, the calibration requires a non-linear optimization approach to fit the tree and can become computationally expensive. Both aforementioned approaches estimate the terminal IRNPD and work backwards in time. Furthermore, the implied binomial trees are arbitrage-free by construction. Another way to construct an implied binomial tree was suggested by Brown and Toft (1999). They use a three-step procedure and get an arbitrage-free, semi-recombining implied binomial tree, which is consistent with the implied volatility surface. However, this method does not use all the available information optimally, since the IRNPD for each maturity date is estimated separately. Moreover, in some cases, the optimization problem can not be solved, since the constraints of the optimization problem can not be satisfied.

Derman and Kani (1994) construct an implied tree assuming that any option value can be interpolated or extrapolated from the prices of actively traded options. Therefore, the resulting implied tree fits the volatility smile in strike and time dimension. Unfortunately, negative transition probabilities can occur and must be replaced by values between zero and one. This may lead to numerical instability of the tree, especially for a large number of time steps. Barle and Cakici (1998) extend the approach by Derman and Kani (1994) to reduce the numerical problems and increase the stability of the algorithm. However, even this approach does not guarantee positive transition probabilities in all nodes.

The main characteristic of implied trinomial trees is that the complete state space for the tree is fixed in advance and only the transition probabilities must be calculated. To construct the implied trinomial tree, missing prices must also be calculated by interpolation or extrapolation of the implied volatility smile. Therefore, the performances of the models depend on the respective interpolation or extrapolation method. Dupire (1994) uses Arrow–Debreu prices implied by the (interpolated/extrapolated) market prices to calculate the transition probabilities, whereas Derman et al. (1996) use the Arrow–Debreu prices of the previous time step calculated by forward induction and the (interpolated/extrapolated) market prices for the next time step. The major drawback of implied trinomial trees is that negative transition probabilities can occur, which is equivalent to the existence of arbitrage opportunities.

Above all, our resulting implied binomial tree is arbitrage-free by construction, and we do not have to worry about negative transition probabilities as in the approaches by Dupire (1994), Derman and Kani (1994), Derman et al. (1996), or Barle and Cakici (1998). Furthermore, our approach neither uses an interpolation or an extrapolation method (e.g. Dupire (1994), Derman and Kani (1994), Derman et al. (1996), or Barle and Cakici (1998)) or a weighting function for the path probabilities, as the one in the approach by Jackwerth (1997). Additionally, the new approach allows for the incorporation of all market information and fits the volatility surface. Moreover, our approach overcomes a deficiency of existing approaches. It is the first approach, to the best of our knowledge, which starts at the current date and works forward to the longest available maturity date using simultaneously backward and forward induction and where the resulting implied binomial or implied multinomial trees are arbitrage-free by construction.

We test our new approach empirically against other approaches on a common set of options. In contrast to other studies (e.g. Jackwerth and Rubinstein (2001) or Dumas et al. (1998)) we focus on a cross-sectional out-of-sample test between two competing derivative markets, instead of testing the out-of-sample time consistency of the different algorithms. Overall, we evaluate six option valuation models: the 'classical' Black–Scholes model; the stochastic volatility model from Heston (1993); two deterministic volatility models, namely the specification of a volatility function and a naive-trader-rule; the Weighted Monte Carlo approach proposed by Avellaneda et al. (2001); and finally our implied tree approach. We implement each model by estimating the implied volatility and other structural parameters from observed option prices collected at the EUREX for each trading day, with a sample period from January 2, 2004 to June 30, 2004.

As already mentioned, our analysis focuses on an out-of-sample pricing performance comparison instead of testing the out-of-sample predictive power for future implied volatilities of the different models. The latter kind of tests are done, e.g. by Bakshi et al. (1997), Dumas et al. (1998), and Jackwerth and Rubinstein (2001). For our study we use market quotes of two competing derivative markets to examine the pricing performance of the models. The key idea of this procedure is to determine the price for a newly issued option. That is why we assume that at one market we observe the correct prices of such newly issued options, whereas the other market is assumed to provide the basis with all currently issued options. Therefore, we take market quotes for DAX index options from the EUREX and use these quotes to calibrate the pricing models. Afterwards, we use the calibrated models to price different types of options that are traded at the European Warrant Exchange (EUWAX) and compare the model prices to the market quotes at EUWAX.

For the overall performance, it turns out that our implied tree approach performs best for all observed option categories. As expected, the Black–Scholes model shows the worst pricing performance. The Weighted Monte Carlo approach performs quite well. The performance of the deterministic volatility models and the Heston model is mixed.

Implied trees work backwards from the maturity date to price different types of options. However, they are inefficient for higher dimensional problems (i.e. problems with more than three state variable or multi-asset or basket options), and they are very difficult to apply to path-dependent options. Simulation techniques are often used here, since they are simple and flexible. The main problem is the calibration of high-dimensional models to generate the sample paths, since often no closed-form pricing formulas are available. Avellaneda et al. (2001) present a new approach for calibrating Monte Carlo simulations to the market prices of benchmark securities. Starting from a given model of market dynamics, they correct misspecifications in the simulation by assigning new 'probability weights' to each simulated path.

Besides this, there exists another common problem in the context of Monte Carlo simulations. For a long time, simulation techniques seemed to be inapplicable to American-style options. This is due to their forward-construction principle and their path-by-path generation. However, since American-style contingent claims are traded in all-important derivative markets, many suggestions have been made to price American-style options by simulation, e.g. Tilley (1993), Andersen (1999), Longstaff and Schwartz (2001), or Ibánez and Zapatero (2004).

In order to price American-style derivatives via Monte Carlo simulation in accordance with given market prices, we propose two new intuitive and efficient valuation methods. Our first approach combines the Weighted Monte Carlo technique by Avellaneda et al. (2001) with the approach by Andersen (1999), while the second method merges the Weighted Monte Carlo method with the Least-Squares Monte Carlo approach by Longstaff and Schwartz (2001). Both approaches are intuitive, easy-to-apply, and computationally efficient. We illustrate the original methods by Andersen (1999) and Longstaff and Schwartz (2001), as well as our extensions, using the example for valuing standard American put options with market data from EUREX and EUWAX. We find that the pricing performance depends on two factors: the chosen 'benchmark' model to generate the Monte Carlo paths and the application of our extensions. In addition, the least-squares approaches perform better than the corresponding threshold approaches, but they are also more time-consuming. Therefore, the user has the choice between a 'quick and dirty' price and more precise prices obtained by using the least-squares approaches. These extensions represent, to the best of our knowledge, the first approaches to a market-conform pricing of American-style options via Monte Carlo simulation.

1.2 Structure of the Thesis

The rest of the thesis is organized as follows:

In the first chapter, we present our new approach to identify the binomial process of the underlying asset price by using a simultaneous backward and forward induction algorithm. After reviewing the related literature, we describe our algorithm to construct an arbitrage-free implied binomial tree. We show possible modifications of the algorithm to speed up the computations. This also allows us to construct arbitrage-free multinomial trees. We compare the model to the existing approaches to emphasize its enhancements. On the basis of a detailed example, we illustrate the flexibility of the model. The appendix contains some technical issues that are important for the simplification of the optimization problem.

Chapter 3 contains the empirical test of different option pricing models based on a common set of options. In detail, we consider the model of Black and Scholes, the model of Heston, naive-trader-rules, deterministic volatility models, implied binomial trees, and weighted Monte Carlo techniques. Each model is described briefly with respect to its assumptions and the associated valuation procedure, as well as the techniques needed for calibration. After introducing the dataset we use the models calibrated to EUREX options to price observed option prices of American call options and knock-out options traded at EUWAX.

In Chapter 4, new methods to value American-style options via Monte Carlo simulations in accordance with given market prices are discussed. After a short introduction to Monte Carlo methods, two new approaches are presented. Specifically, we combine the weighted Monte Carlo approach proposed by Avellaneda et al. (2001) with the approach proposed by Andersen (1999), as well as with the Least-Squares Monte Carlo (LSM) approach suggested by Longstaff and Schwartz (2001). We show the effect of weighting Monte Carlo paths to reproduce market prices correctly in the ideal case of a completely specified model. Afterwards, we compare the effect with market data and illustrate the original techniques and our extensions based on the valuation of standard American put options.

In the final section of each chapter, the main results of the chapter are summarized and areas for future research are pointed out. A more detailed description of the chapter contents can be found in the first section of each chapter.

References

Aït-Sahalia, Y. and A.W. Lo (1998): Nonparametric Estimation of State-Price Densities Implicit in Financial Asset Prices, *The Journal of Finance*,
53(2):499–547.

Amin, K. and R. Jarrow (1992): Pricing Options on Risky Assets in a Stochastic Interest Rate Economy, *Mathematical Finance*, 2(4):217–237.

Andersen, L. (1999): A Simple Approach to the Pricing of Bermudan Swaptions in the Multi-Factor LIBOR Market Model, *The Journal of Computational Finance*, 3(2):5–32.

Avellaneda, M., R. Buff, C. Friedman, N. Grandchamp, L. Kruk, and J. Newman (2001): Weighted Monte Carlo: A New Technique for Calibrating Asset-Pricing Models, *International Journal of Theoretical and Applied Finance*, 4(1):91–119.

Bakshi, G., C. Cao, and Z. Chen (1997): Empirical Performance of Alternative Option Pricing Models, *The Journal of Finance*, 52(5):2003–2049.

Bakshi, G. and Z. Chen (1997): An Alternative Valuation Model for Contingent Claims, *Journal of Financial Economics*, 44(1):123–165.

Barle, S. and N. Cakici (1998): How to Grow a Smiling Tree, *Journal of Financial Engineering*, 7(2):127–146.

Bates, D. (1991): The Crash of '87: Was It Expected? The Evidence from Options Markets, *The Journal of Finance*, 46(3):1009–1044.

Bates, D. (1996): Jumps and Stochastic Volatility: Exchange Rate Processes Implicit in Deutsche Mark Options, *The Review of Financial Studies*, 9(1):69–108.

Belledin, M. and C. Schlag (1999): An Empirical Comparison of Alternative Stochastic Volatility Models, *Working Paper Series Finance & Accounting, Fachbereich Wirtschaftswissenschaften*, Goethe–University, Frankfurt am Main.

Black, F. and M. Scholes (1973): The Valuation of Options and Corporate Liabilities, *Journal of Political Economy*, 81(3):637–654.

Brown, G. and K.B. Toft (1999): Constructing Binomial Trees from Multiple Implied Probability Distributions, *The Journal of Derivatives*, 7(2):83–100.

Cox, J.C., S.A. Ross, and M. Rubinstein (1979): Option Pricing: A Simplified Approach, *Journal of Financial Economics*, 7:229–263.

Das, S.R. and R.K. Sundaram (1999): Of Smiles and Smirks: A Term Structure Perspective, *The Journal of Financial and Quantitative Analysis*, 34(2):211–239.

Derman, E. and I. Kani (1994): Riding on a Smile, *RISK*, 7(2):32–39.

Derman, E., I. Kani, and N. Chriss (1996): Implied Trinomial Trees of the Volatility Smile, *The Journal of Derivatives*, 3(4):7–22.

Dumas, B., J. Fleming, and R.E. Whaley (1998): Implied Volatility Functions: Empirical Tests, *The Journal of Finance*, 53(6):2059–2106.

Dupire, B. (1994): Pricing with a Smile, *RISK*, 7(1):18–20.

Harrison, J.M. and D.M. Kreps (1979): Martingales and Arbitrage in Multiperiod Securities Markets, *Journal of Economic Theory*, 20(3):381–408.

Heston, S. (1993): A Closed-Form Solution for Options with Stochastic Volatility with Applications to Bond and Currency Options, *The Review of Financial Studies*, 6(2):327–343.

Hull, J.C. and A. White (1987): The Pricing of Options on Assets with Stochastic Volatilities, *The Journal of Finance*, 42(2):281–300.

Ibáñez, A. and F. Zapatero (2004): Monte Carlo Valuation of American Options Through Computation of the Optimal Exercise Frontier, *The Journal of Financial and Quantitative Analysis*, 39(2):253–275.

Jackwerth, J.C. (1997): Generalized Binomial Trees, *The Journal of Derivatives*, 5(2):7–17.

Jackwerth, J.C. (1999): Option Implied Risk-Neutral Distributions and Implied Binomial Trees: A Literature Review, *The Journal of Derivatives*, 7(2):66–81.

Jackwerth, J.C. and M. Rubinstein (1996): Recovering Probability Distributions from Option Prices, *The Journal of Finance*, 51(5):1611–1631.

Jackwerth, J.C. and M. Rubinstein (2001): Recovering Stochastic Processes from Option Prices, *Working Paper*, London Business School.

Longstaff, F. and E. Schwartz (2001): Valuing American Options by Simulation: A Simple Least-Squares Approach, *The Review of Financial Studies*, 14(1):113–147.

Merton, R.C. (1973): Theory of Rational Option Pricing, *Bell Journal of Economics and Management*, 4:141–183.

Merton, R.C. (1976): Option Pricing when Underlying Stock Returns are Discontinuous, *Journal of Financial Economics*, 3:125–144.

Rubinstein, M. (1994): Implied Binomial Trees, *The Journal of Finance*, 49(3):771–818.

Schöbel, R. and J. Zhu (1999): Stochastic Volatility With an Ornstein-Uhlenbeck Process: An Extension, *European Finance Review*, 3(1):23–46.

Scott, L.O. (1997): Pricing Stock Options in a Jump-Diffusion Model with Stochastic Volatility and Interest Rates: Applications of Fourier Inversion Methods, *Mathematical Finance*, 7(4):413–426.

Skiadopoulos, G. (2001): Volatility Smile Consistent Option Models: A Survey, *International Journal of Theoretical and Applied Finance*, 4(3):403–437.

Tilley, J.A. (1993): Valuing American Options in a Path-Simulation Model, *Transactions of Society of Actuaries*, 45:499–520.

Construction of Arbitrage-Free Implied Trees: A New Approach

2.1 Introduction

It is well known that the performance of hedges for exotic options can be improved by consistent valuation of exotic and plain-vanilla options. However, the market-conform pricing of non-standard options is quite challenging. A popular set of approaches to handle this problem are the so-called 'implied tree models', which are extensions of the seminal binomial model developed by Cox et al. (1979). They are called *implied* trees, because they are inferred from market prices of traded options. Therefore, they are consistent with the volatility smile and can be viewed as a discretization of generalized one-dimensional diffusions in which the volatility parameter is allowed to be a deterministic function of both time and asset price. To build a consistent risk-neutral price process of the underlying, implied trees take market prices of liquid standard options as given. These options are then used to infer information about the data-generating process. In order to construct an implied tree, we need to know the state space (i.e. the asset prices at the different nodes of the tree), as well as the transition probabilities. Approaches to extract this information from the data are proposed by Rubinstein (1994), Dupire (1994), Derman and Kani (1994), Jackwerth (1997) or Brown and Toft (1999).

We propose a new, more powerful and flexible method to construct arbitrage-free implied binomial trees via simultaneous backward and forward induction based on the approach by Brown and Toft (1999). As the output of our procedure, we get an arbitrage-free, risk-neutral implied binomial tree, which is consistent with the term structure of implied volatilities and also with the implied volatility smile. The implied risk-neutral probability distributions (IRNPDs) for later maturity dates are an endogenous result of the model and take the IRNPDs of the prior maturity dates into account. This extends the approach by Brown and Toft (1999), where the IRNPDs of each maturity date are estimated separately and afterwards a set of conditional distributions is determined to connect the IRNPDs of two successive maturity dates. In contrast to other approaches, e.g. Dupire (1994), Derman and Kani (1994), Derman et al. (1996), or Barle and Cakici (1998), our approach neither uses an interpolation or an extrapolation method nor a weighting function for the path probabilities, as in the approach by Jackwerth (1997). We only have to solve an optimization problem, which can be reduced to the case of a quadratic objective function with linear inequality constraints. This type of problem is well-studied and can easily be solved with standard routines. Moreover, in our approach, the user only has to specify the state space at each maturity date. No prior guesses for the IRNPDs and transition probabilities are necessary, which reduces the risk of making 'wrong' choices. Above all, the resulting implied binomial tree is arbitrage-free by construction, and we do not have to worry about negative transition probabilities as in the approaches by Dupire (1994), Derman and Kani (1994), Derman et al. (1996), or Barle and Cakici (1998).

Finally, our method can also be used to construct arbitrage-free, risk-neutral implied multinomial trees. These multinomial trees allow us to construct hedge-portfolios of the money market account and the underlying asset and traded options using standard replication arguments. To realize this method, the number of basis assets has to be equal to the number of branches of the multinomial tree. Moreover, the multinomial setting can be used to calibrate models with more than one state variable, e.g. the underlying price process and stochastic volatility.

The remainder of the chapter is structured as follows. The next section summarizes the key ideas of related papers. Section 2.3 describes our algorithm to construct an implied binomial tree consistent with given market prices. A detailed example is presented in Sect. 2.4, and Sect. 2.5 concludes with a brief summary and a discussion of issues for further research.

2.2 Related Literature

Since the approaches suggested by Rubinstein (1994), Dupire (1994), Derman and Kani (1994), Derman et al. (1996), Jackwerth and Rubinstein (1996), Jackwerth (1997), Barle and Cakici (1998) and Brown and Toft (1999) are closely related to this chapter, they are described briefly in this section. For more detailed surveys see Jackwerth (1999) or Skiadopoulos (2001).

The existing approaches to construct implied trees can be divided into three classes. First, implied binomial trees are often constructed by only using backward induction. These trees fit either the volatility smile (Rubinstein (1994)) or both the volatility smile and the time dimension of implied volatilities (Jackwerth (1997)). The second class also consists of implied binomial trees, but these are constructed by using backward and forward induction (Derman and Kani (1994), Barle and Cakici (1998), Brown and Toft (1999)). The third class contains implied trinomial trees, which are built by using simultaneous backward and forward induction (Dupire (1994), Derman et al. (1996)). Both the second and the third class fit the smile and the term structure of implied volatilities. The main difference between the latter two classes is that the complete state space of implied trinomial trees is fixed in advance, and, therefore, only the transition probabilities have to be specified.

The main characteristic of the first class is that the IRNPD at the terminal date of the tree is estimated directly from the option prices without interpolating missing prices. This is done by solving an optimization problem. Then backward induction is used to construct an implied tree, which matches this terminal IRNPD. According to the approach suggested by Rubinstein (1994), the terminal IRNPD is chosen to be close to a prior guess, subject to some constraints. Rubinstein's main assumption is that all paths that lead to the same terminal node have the same risk-neutral probability. This construction principle guarantees an arbitrage-free implied binomial tree. However, there are two major drawbacks to Rubinstein's method. First, the implied binomial tree fits the strike dimension of the volatility smile only at one single maturity date and neglects information for traded options with shorter or longer maturities. This leads to the second problem, namely that implied binomial trees constructed for two different maturity dates are not necessarily consistent for the overlapping period.

Jackwerth and Rubinstein (1996) test other objective functions for the optimization problem and suggest the use of the 'smoothness criterion' to recover the IRNPD from given market prices. The smoothest distribution is the one that minimizes the curvature of the implied probability distribution. The advantage is that no prior guess about the terminal IRNPD is necessary. Furthermore, the authors propose that the distance between the option prices under the IRNPD and the midpoint market prices should be minimized instead of demanding that the calculated option prices under the IRNPD fall between the respective bid and ask prices.

To overcome the problems of the Rubinstein (1994) approach, Jackwerth (1997) develops a generalized implied binomial tree by relaxing the assumption that all paths ending up in the same terminal node have the same probability. This is done by introducing a piecewise-linear weight function and by using nodal probabilities instead of path probabilities. This generalization allows for the incorporation of all the information and to fit the volatility smile in the strike and time dimensions. However, the calibration requires a non-linear optimization approach to fit the tree and can become computationally expensive, since in each optimization step the complete tree must be constructed to incorporate options with different maturity dates. Therefore, the value and the cash-flow of each option with shorter maturity must be recalculated, since the state space for shorter maturities changes in each optimization step.

The first approach belonging to the second class was suggested by Derman and Kani (1994). To build the implied tree, they assume that any option price or, more precisely, any implied volatility can be interpolated or extrapolated from the prices of actively traded options. Thereby, the asset price for each node for a new time step and the transition probability to reach this node are determined by the (interpolated or extrapolated) market price of the European plain-vanilla option with a strike price equal to asset price of the previous step and with an expiration at the end of the next time step. Therefore, the resulting tree fits the volatility smile in the strike and time dimensions. Unfortunately, negative transition probabilities can occur and must be replaced by positive values. This may lead to the numerical instability of the tree, especially for a large number of time steps.

Barle and Cakici (1998) extend the approach by Derman and Kani (1994) to reduce the numerical problems and increase the stability of the algorithm. However, this approach as well does not guarantee positive transition probabilities in all nodes. Barle and Cakici (1998) note that with increasing interest rate and/or increasing slope coefficients of the volatility smile negative transition probabilities are encountered even more often than with the modified method.

Another way to construct an implied binomial tree by using backward and forward induction was suggested by Brown and Toft (1999). They use a three-step procedure. First, they estimate the IRNPD for each maturity date separately by assuming that a volatility function for each maturity date exists. Afterwards, they estimate the conditional implied probability distributions between two maturity dates. Finally, they use the algorithm from Rubinstein (1994) to construct implied binomial sub-trees. As a result, they obtain an arbitrage-free, semi-recombining implied binomial tree, which is consistent with the implied volatilities in the strike and time dimensions. The resulting binomial process is only semi-recombining, because the complete tree shares the same nodes at different maturity dates, but they will not generally match up between maturity dates. However, one disadvantage of this method is that not all of the available information is used optimally, since the implied risk-neutral terminal distribution for each maturity date is estimated separately. This leads to a further problem, namely, that in some cases no conditional implied distribution satisfies the constraints.

The main characteristic of implied trinomial trees, the third class of approaches, is that the complete state space for the tree is fixed in advance and only the transition probabilities must be calculated. Additionally, the tree is built by simultaneously using backward and forward induction to satisfy the forward condition[1] of the underlying process and to fit the market prices of plain-vanilla European-style options, which are used as benchmark instruments. Therefore, prices of plain-vanilla European-style options with a strike price equal to the asset price of the previous step and expiration at the end of the next time step are needed for construction. However, in real markets a complete set over all strike prices and maturity dates is rarely available. Hence, missing prices must be calculated by interpolation or extrapolation of the implied volatility smile based on the market prices. Therefore, the performance of the model depends on the respective interpolation or extrapolation method.

Dupire (1994) uses Arrow–Debreu prices implied by the (interpolated/extrapolated) market prices to calculate the transition probabilities. However, Derman et al. (1996) compute the transition probabilities using the Arrow–Debreu prices of the previous time step calculated by forward induction and the (interpolated or extrapolated) market prices for the next time step. The possibility of negative transition probabilities is the major drawback of these approaches. This problem can be mitigated by choosing an adequate state space, which depends on the structure of the volatility smile. Derman et al. (1996) discuss several methods to choose an adequate state space as well as ways to replace negative probabilities.

---

[1] The forward condition demands that the expected value, one period later, of the stock at any node in the tree must be its known forward price, which guarantees that risk-neutrality is satisfied.

Note that implied binomial and trinomial trees constructed by interpolating or extrapolating the volatility smile have the disadvantage of potentially negative transition probabilities. If, on the other hand, negative probabilities are replaced by positive values following a certain algorithm, information about the volatility smile gets lost. How severe this problem is also depends on the method used to fill in missing option prices. In general, one can say that with an increasing number of replaced negative probabilities the quality of the results will decrease.

2.3 Constructing Implied Trees

2.3.1 The Model

We will first describe our new approach in detail. Afterwards, possible simplifications are presented to reduce the complexity and the computation time of the approach. Finally, we compare our approach to the existing models reviewed in the previous section.

We take market prices of liquid-traded European plain-vanilla options as given to fit the IRNPD for each option maturity date $T_1 < T_2 < ... < T_n$, where n is the number of available maturity dates. Afterwards, we use the method proposed by Rubinstein (1994) to construct a semi-recombining implied binomial tree. The resulting tree is arbitrage-free by construction and can be used to price and hedge a wide range of standard and exotic options.

The node numbers of the binomial tree are ordered from lowest to highest, where the lowest node in each time step is labelled '1'. We use equal time steps Δt to construct the tree, i.e. $\Delta t = \frac{T_l}{N_l - 1} = \frac{T_{l+1}}{N_{l+1} - 1}$ for $l = 1, ..., n-1$, where $N_l - 1$ is the number of time steps associated with maturity date T_l. To simplify matters, we assume a constant interest rate r for each time step. Note that this assumption is not crucial, since the term structure of interest rates can easily be integrated in our model. Without loss of generality, we focus on a non-dividend paying underlying asset. Moreover, in accordance with Rubinstein (1994), we assume that all paths that lead to the same node have the same risk-neutral probability.

The algorithm consists of three steps. In Step 1, we arbitrarily set the state space for each maturity date in advance. For example, this can be done by estimating the implied volatility of all traded benchmark options, which minimizes the sum of squared pricing errors. Based on this, the state space can be determined by using the method from Cox et al. (1979) to construct a standard binomial tree with constant volatility and equal time steps Δt. In Step 2, we estimate the transition probabilities for the sub-trees between all maturity dates. For the first maturity date T_1, we apply the algorithm proposed by Rubinstein (1994) in combination with the extension of Jackwerth and Rubinstein (1996) to recover the implied risk-neutral probability distribution (IRNPD). This means that we identify the IRNPD by minimizing the 'smoothness criterion' and the sum of squared pricing errors of traded benchmark instruments under this distribution. We prefer this criterion, since it allows us to simplify the resulting optimization problem to a well-known structure. Furthermore, Jackwerth and Rubinstein (1996) point out that the choice of method does not really matter much because most criteria back out virtually to the same IRNPD, as long as there is a sufficient number of options, usually more than 10. In Step 3, we use the estimated transition probabilities from Step 2 to construct binomial sub-trees between the different maturity dates.

After having estimated the IRNPD of the first maturity date T_1, we have to estimate the implied risk-neutral probabilities for all maturity dates $T_2, ..., T_n$ based on the known IRNPDs of all previous maturity dates. Suppose we have estimated the IRNPDs up to the $l-th$ maturity date. Figure 2.1 shows the structure of the tree at the $l-th$ maturity date T_l with N_l nodes. For each node (T_l, j) we know the asset price $S_{T_l,j}$, the implied risk-neutral probability $Q_{T_l,j}$ to reach this node, and the asset prices $S_{T_{l+1},k}$ for all nodes $k = 1, ..., N_{l+1}$ ($N_{l+1} > N_l$) at the next maturity date, since we chose the complete state space in advance. This structure can be generated by standard binomial trees with $N_l - 1$ time steps to the $l-th$ maturity date. Then we have $\Delta N_l \equiv N_{l+1} - N_l$ time steps between the two maturity dates. This implies that each node (T_l, j) has $M_l \equiv N_{l+1} - N_l + 1$ reachable nodes at the next maturity date. We call $q_{j,m}$ the transition probability to reach node $(T_{l+1}, j+m-1)$ from node (T_l, j) for $m = 1, ..., M$. Note that emanating from each single node (T_l, j) at maturity date T_l the corresponding transition probabilities $q_{j,m}$ represent a structure that is similar to the structure of the implied binomial tree suggested by Rubinstein (1994). Hence, we refer to these transition probabilities $q_{j,m}$ emanating from a special node (T_l, j) as a sub-tree.

The implied risk-neutral probability $Q_{T_{l+1},k}$ to reach node (T_{l+1}, k) is the sum of all path probabilities entering that node, i.e.

$$Q_{T_{l+1},k} = \sum_{j \geq 1, m \geq 1, k=j+m-1} q_{j,m} Q_{T_l,j} \qquad (2.1)$$

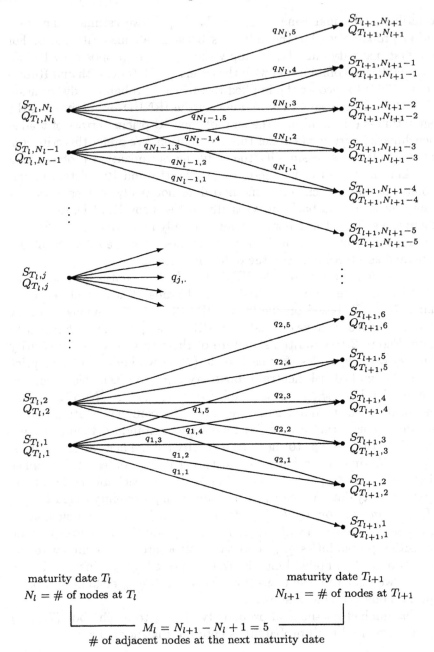

Fig. 2.1. Construction of the implied tree for the T_{l+1} maturity date. To simplify the demonstration we set $M_l = 5$.

Therefore, we have to determine all transition probabilities $q_{.,.}$ to calculate the IRNPD for the maturity date T_{l+1}. We use the 'smoothness criterion' defined by Jackwerth and Rubinstein (1996) to compute transition probabilities between the maturity dates and the probability distribution for the next maturity date. Furthermore, under the resulting probability distribution, the squared pricing error is minimized, given the market prices of the benchmark instruments at the next maturity date. For each maturity date T_{l+1}, this leads to the following quadratic optimization problem:

$$
\min_{q_{j,m}} \alpha \sum_{j=1}^{N_l} \sum_{m=1}^{M_l} \left(q_{j,m-1} - 2q_{j,m} + q_{j,m+1}\right)^2
$$
$$
+ \beta \sum_{k=1}^{N_{l+1}} \left(Q_{T_{l+1},k-1} - 2Q_{T_{l+1},k} + Q_{T_{l+1},k+1}\right)^2 \tag{2.2}
$$
$$
+ \gamma \sum_{i=1}^{O} \frac{1}{w_i} \left(\sum_{k=1}^{N_{l+1}} Q_{T_{l+1},k} CF_{k,i} - CM_i \exp\left(r N_{l+1} \Delta t\right)\right)^2
$$

subject to

$$
\sum_{m=1}^{M_l} q_{j,m} = 1 \quad \text{for } j = 1, ..., N_l, \tag{2.3}
$$

$$
q_{j,m} \geq 0 \quad \text{for } j = 1, ..., N_l, \ m = 1, ..., M_l, \tag{2.4}
$$

$$
\sum_{m=1}^{M_l} q_{j,m} S_{T_{l+1},j+m-1} = S_{T_l,j} \exp\left(r \Delta_l \Delta t\right) \quad \text{for } j = 1, ..., N_l, \tag{2.5}
$$

where $CF_{k,i}$ is the relative cash flow of option i in state k at maturity date T_{l+1}, CM_i is the corresponding relative market price[2], w_i is a weighting factor for each benchmark option, and O is the number of traded benchmark options. We use relative prices and cash flows in order to ensure an equal impact of all three terms on the objective function, since after normalization the values of the third term in (2.2) are also between zero and one. Notice that in the limit $(w_i \rightarrow 0)$ we obtain an exact matching of the market prices. We propose to use $w_i = CM_i$ so that out-of-the-money options get relatively higher weights as compared with in-the-money options, since the former are supposed to contain more information about the volatility smile. Furthermore, we introduce the penalty parameters α, β and γ to weigh the smoothness criterion of the transition probabilities, the smoothness criterion of the IRNPD, and the accuracy of matching the market prices. The choice of the penalty

---

[2] Relative cash flow and relative market price are the cash flow and market price, respectively, normalized by the current underlying price S_{T_0}.

parameters depends on the desired trade-off between the different objectives. This means, if we want to have a smooth distribution at the maturity dates with correct pricing rather than smooth transition probabilities, we have to choose $\alpha < \beta < \gamma$. Numerical tests indicate that the penalty parameters $\alpha = \frac{1}{N_l}$, $\beta = 1$ and $\gamma = \frac{1000}{N_{l+1}}$ are advisable for the objectives described above.

The restrictions (2.3)–(2.5) of the optimization problem are typical economic constraints for implied trees. Constraint (2.3) forces the transition probabilities for each sub-tree to sum up to one, whereas constraint (2.4) guarantees that all transition probabilities are greater than or equal to zero. Finally, the forward condition (2.5) ensures risk-neutral pricing. These restrictions are needed to get an arbitrage-free tree.

Rewriting the problem in matrix notation shows that it is equivalent to a well-studied quadratic programming problem:

$$\min_{\mathbf{q}} \quad \mathbf{q'Hq} + \mathbf{b'q} + c \tag{2.6}$$

subject to

$$\mathbf{q} \geq \mathbf{0}_{N_l M_l} \tag{2.7}$$

$$\mathbf{Aq} = \mathbf{y} \tag{2.8}$$

with

$$\mathbf{H} := \alpha \hat{\mathbf{B}}' \hat{\mathbf{B}} + \hat{\mathbf{Q}}'(\beta \mathbf{B}'_{N_{l+1}} \mathbf{B}_{N_{l+1}} + \gamma \mathbf{C}'_F \mathbf{W} \mathbf{C}_F) \hat{\mathbf{Q}} \tag{2.9}$$

$$\mathbf{b} := -2\gamma (\mathbf{C}_F \hat{\mathbf{Q}})' \mathbf{W} \mathbf{C}_M \tag{2.10}$$

$$c := \gamma \mathbf{C}'_M \mathbf{W} \mathbf{C}_M \tag{2.11}$$

and where the vectors and matrices are given in detail in the appendix. The Hessian matrix $\mathbf{H}$ is symmetric, which makes the problem perfectly suitable for standard optimization routines. Due to the fact that $\hat{\mathbf{Q}}$, $\hat{\mathbf{B}}$ and $\mathbf{B}_{N_{l+1}}$, are sparse matrices, the calculation of $\mathbf{H}$ and $\mathbf{b}$ is rather easy.

After the optimization, we use the Rubinstein (1994) algorithm to convert the calculated transition probabilities $q_{j,m}$ into N_l different implied binomial sub-trees each emanating from node (T_l, j). Hence, the resulting implied binomial tree is not recombining in the traditional sense, since the nodes of the implied sub-trees generally do not match up between different maturity dates. We call this occurrence semi-recombining in accordance with Brown and Toft (1999). However, by construction, we have the same state space at each maturity date with an IRNPD matching the prices of all traded benchmark options in the sense of least-squares.

After repeating the algorithm up to the last maturity date, we get a semi-recombining risk-neutral and arbitrage-free implied binomial tree to price plain-vanilla and non-standard options. We are also able to price some types of path-dependent options via Monte Carlo simulation by randomly sampling paths throughout the tree.

2.3.2 Possible Simplifications

The setting of our algorithm leads to an exponential growth of unknowns in N_l. Therefore, a reduction of the complexity of the problem seems desirable. First, one could reduce the number of sub-trees between two maturity dates. This method was proposed by Brown and Toft (1999). Second, it would also be possible to transform the problem into a multinomial tree setting by choosing $M_l \equiv M$, in combination with an interpolation or approximation of the volatility smile. Here, M is the number of branches in the multinomial tree, where, for example, $M = 3$ corresponds to a trinomial tree.

For the first method, we assume that adjacent nodes have the same transition probabilities, i.e. we relax the restriction of a state and time dependent transition probability for each node at the $l-th$ maturity date. This means that instead of estimating the transition probabilities for N_l sub-trees, one for each node at the $l-th$ maturity date, we only estimate the transition probabilities for $\bar{N}_l < N_l$ sub-trees. In accordance with Brown and Toft (1999) we set

$$q_{i_1,m} = q_{i_2,m} \quad \forall i_1, i_2 \in (d_j, ..., D_j) \quad \text{for } j = 1, ..., \bar{N}_l, \tag{2.12}$$

where j indexes distinct transition probabilities and d_j and D_j denote the lowest and highest node index i at time T_l, to which we assign the same transition probability $q_{j,m}$. For example, imagine that the transition probabilities for three adjacent nodes are forced to be equal. Furthermore, suppose there are nine nodes at the $l-th$ maturity date. Then we have $\bar{N}_l = 3$, $d_1 = 1$, $D_1 = 3$, $d_2 = 4$, $D_2 = 6$, $d_3 = 7$, and $D_3 = 9$. This setting produces a reduction of the linear constraints from $2N_l$ to $2\bar{N}_l$, and the number of unknowns is reduced from $N_l M_l$ to $\bar{N}_l M_l$. Consequently, in contrast to Brown and Toft (1999), $\bar{N}_l$ can be arbitrarily chosen as long as $M_l > 2$. Otherwise, if $M_l \le 2$, then there are not enough degrees of freedom to satisfy all the restrictions on the transition probabilities $q_{j,m}$. This changes some matrices and vectors in the optimization problem (2.6)–(2.11), but the general structure remains unaffected.[3]

For the second simplification, we assume that there is a volatility function $\sigma(K, T)$ for all possible strike prices K and maturity dates T. Given this, we can use interpolation or approximation methods to fit the volatility function. This allows us to calculate prices for European-style options with arbitrary K and T. Now, we can compute transition probabilities between arbitrary points in time rather than only between two option maturity dates. Consequently, we are able to construct multinomial trees. To illustrate this, we set $M_l \equiv M = 3$, which corresponds to a trinomial tree. Then the structure of the tree in Fig. 2.1 is simplified to the common structure of trinomial trees. Given the function $\sigma(K, T)$, we use the Black–Scholes formula to price European options maturing at the next time step $(t+1)$ with strike prices corresponding to the strike prices of the traded options maturing next at $T_l > t + 1$. Consequently, we get a 'synthetic' set of benchmark options to run our optimization. As an advantage, we do not need Step 3 to construct implied sub-trees anymore. Unfortunately, we pay for this with the assumption that a volatility function exists. By applying this simplification, we only have $M \times N_l$ remaining variables to estimate. The number of linear constraints does not change, and we have $(M - 2) \times N_l$ degrees of freedom in our optimization problem. Due to the structure of our problem, the computational time to build the complete implied multinomial tree is manageable for reasonable values of N_l and small values of M. As a result, we get a risk-neutral, recombining, and arbitrage-free implied multinomial tree, which fits the volatility smile in the time and strike dimensions.

2.3.3 Comparison to Existing Approaches

The proposed approach offers some advantages over existing models. One main feature of our algorithm compared to all others is that the user only has to specify the state space at each maturity date. All other parameters of the tree are an output of our model. On the other hand, the main disadvantage is the growing computational time if the number of unknown variables increases. However, due to the rate of development

---

[3] For $\mathbf{q}$, $\hat{\mathbf{B}}$, and $\mathbf{A}$ these modifications should be obvious. Only for $\hat{\mathbf{Q}}$, the changes are somewhat more complex. $\hat{\mathbf{Q}}$ has N_l sub-matrices, each of size $N_{l+1} \times M_l$. Thereby, each sub-matrix corresponds to one node at the $l-th$ maturity date. To obtain the modified sub-matrices, sum up the corresponding sub-matrices of state i with $i \in (d_j, ..., D_j)$.

of more powerful computers, this problem should become less and less important.

Let us now compare our model to other approaches in more detail. In contrast to the approach suggested by Brown and Toft (1999), we do not estimate IRNPDs separately, but we get the IRNPD of the next maturity date as an output of our model, which leads to a reduction of restrictions and ensures that the IRNPD at the next maturity date integrates to one. The forward condition between two maturity dates is also always satisfied by the construction. In the approach of Brown and Toft (1999), the estimation of the implied conditional distributions sometimes fails, since the IRNPDs are determined for each maturity date in advance. We do not estimate the IRNPD for each maturity date separately. Instead, our algorithm yields the IRNPD for the next maturity date as an endogenous outcome of our optimization procedure. This has the additional advantage of using all the information contained in market prices about the asset price process and the volatility smile up to the respective maturity date. We use the 'smoothness criterion' introduced by Jackwerth and Rubinstein (1996), so that we do not have to set any priors for the IRNPD and the transition probabilities in each sub-tree. The only problem left to the user is to fix an adequate state space.

The classical approach to construct implied binomial trees developed by Rubinstein (1994) represents a special case of our model. It emerges if only options for one maturity date are traded. The advantage of our model over Rubinstein's approach is the ability to fit the complete volatility smile in strike and time dimensions. This problem has already been solved by Jackwerth (1997). However, his method requires the user to specify some kink points in the weighting function for calculating the nodal probabilities. Furthermore, the complete tree must be calculated in each optimization step to match all market prices simultaneously. This can become computationally time-consuming, especially when many benchmark options are available.

In the previous section, we stated that our method can also be used to construct implied multinomial trees, which included the common class of implied trinomial trees as a special case. For this purpose, we must assume the existence of a volatility function $\sigma(K, T)$. This is in line with proposals by Dupire (1994) and Derman et al. (1996). The main advantage of the procedure developed in this chapter over these approaches is that the resulting trinomial tree is arbitrage-free by construction. Furthermore, we only generate a 'synthetic' sample of benchmark instruments using the volatility function $\sigma(K, T)$ and the strike prices of the 'traded' benchmark instruments maturing at $T_l > t + 1$ to estimate the transition probabilities. This is in contrast to Dupire (1994) and Derman et al. (1996), which rely on the existence of option prices for all possible strike prices and maturities corresponding to each node in the tree to determine the transition probabilities. This leads to a greater flexibility as compared with the existing approaches, since our 'synthetic' sample is generally much smaller than a sample of options with strike prices equal to the state price of each node.

2.4 Example

Let us now illustrate our model via a detailed example. We want to demonstrate the technique and compare the IRNPD of our method to the IRNPD obtained from the method developed by Jackwerth and Rubinstein (1996). The latter is also one of the methods recommended by Brown and Toft (1999) to identify the IRNPDs, which they use as the input for their algorithm to estimate the conditional probability distributions.

We take all bid and ask quotes for the December, January, February, and March DAX index option quoted at EUREX on December 1st, 2003 at 12:00 P.M. CET. Their corresponding times to maturity are 21, 49, 84, and 112 days, respectively. As a proxy for the risk-free rate we use the 3-month EURIBOR, which was 2.16 % p.a. on that day. Since the DAX is a performance index with dividend reinvestment, we do not need to pay attention to the payout rate. The level of the DAX index is taken from XETRA, which provides the best bid and offer prices. The midpoint of the bid and ask quotes was 3799.505 at the time of sampling.

In a next step, general arbitrage violations have to be eliminated from the data, since we want to estimate risk-neutral probability distributions. Therefore, we apply the same method as Jackwerth and Rubinstein (1996) to sort out arbitrage violations in the data. Hence, two options are sorted out so that the final sample consists of 67 options – 16, 14, 17, and 20 for the different maturity dates. Afterwards, we use the midpoint of the bid-ask quotes to compute the implied volatilities by inverting the Black–Scholes formula. The scatter plot of the implied volatilities for each maturity is given in Fig. 2.2. Note that these structures are typical in equity markets and similar to those documented by Rubinstein (1994), Bakshi et al. (1997), Dumas et al. (1998), and Aït-Sahalia and Lo (1998) for S&P 500 index options.

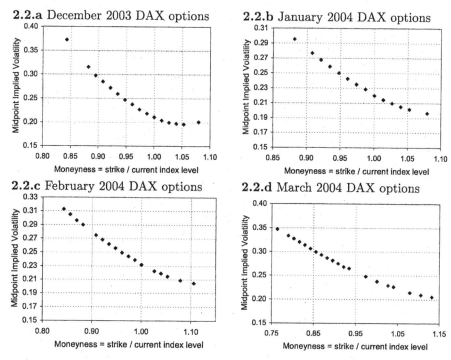

2.2.a December 2003 DAX options

2.2.b January 2004 DAX options

2.2.c February 2004 DAX options

2.2.d March 2004 DAX options

Fig. 2.2. Implied volatility structure for different maturity dates of DAX index options on December 1st, 2003.

The time step is $\Delta t = \frac{1}{365}$. To specify the size of an upward move of the underlying during Δt, we use the volatility under which the sum of squared pricing errors of all traded options is minimized. This results in an implied volatility estimate of 0.23995. The weighting factors are specified as $w_i = CM_i$ and as penalty parameters we use $\alpha = \frac{1}{N_l}$, $\beta = 1$, and $\gamma = \frac{1000}{N_{l+1}}$ (see Sect. 2.3.1). Note that the Jackwerth and Rubinstein (1996) procedure corresponds to our optimization problem (2.2)–(2.5) with $\alpha = 0$, $\hat{Q} = E$, and $N_l = 1$. As the benchmark distribution, we use the log-normal distribution (denoted by 'LOGNORMAL') with the same volatility parameter that was used to specify the size of an upward movement of the underlying. We calculate one sub-tree for each node between the first three maturity dates, whereas between February and March we apply the first simplification described in Sect. 2.3.2 and use the same transition probabilities for three adjacent nodes, i.e. we allow for $\bar{N}_l = 29$ distinct transition probabilities. This is done to reduce the number of unknown parameters.

To specify the state space at each maturity date, we first use the standard method proposed by Cox et al. (1979). This method to specify the state space leads to a high number of nodes with zero probability. This problem becomes apparent in Fig. 2.3, where we have estimated the IRNPD for each maturity date using the Jackwerth and Rubinstein (1996) algorithm ('SMOOTHNESS') and our algorithm ('OPTIMIZATION'), respectively. At this point, it is important to recall the differences between both approaches. Jackwerth and Rubinstein (1996) estimate the IRNPD separately for each maturity date and independently of the other maturity dates, whereas our algorithm estimates the IRNPD by forward induction depending on the IRNPD of former maturity dates.

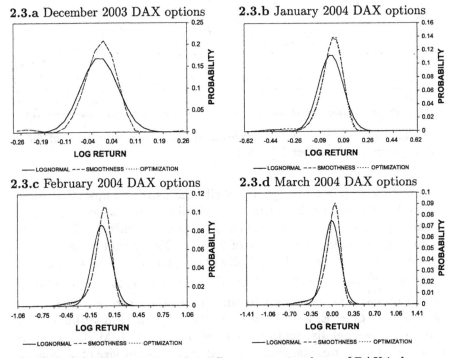

Fig. 2.3. Probability densities for different maturity dates of DAX index options on December 1st, 2003 with state-space specification similar to standard binomial trees.

Comparing the IRNPD estimated with the Jackwerth and Rubinstein (1996) procedure to the IRNPD recovered using our forward optimization algorithm reveals some interesting results. It can easily be seen that there are only small differences between the IRNPDs. In fact, sometimes they are hardly visible. This confirms that our forward induction algorithm, which depends on IRNPDs of former maturity dates, works quite well, since the differences compared to separate estimation of the IRNPDs for single maturity dates are small. Furthermore, one should note that the IRNPDs estimated by separate application of the 'smoothness criterion' can be used as input in the Brown and Toft (1999) algorithm, which might lead to the conclusion that our method is not much different. However, our method is less restrictive and more flexible than the approach by Brown and Toft (1999). Additionally, the optimization problem in our approach is reduced to a standard quadratic optimization problem.

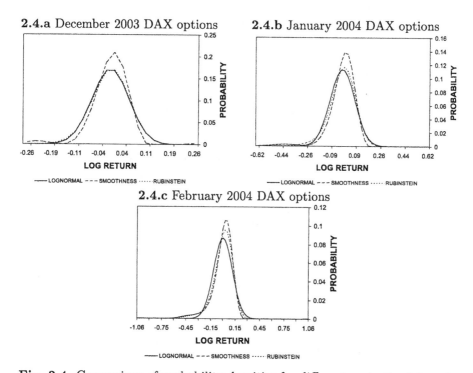

2.4.a December 2003 DAX options

2.4.b January 2004 DAX options

2.4.c February 2004 DAX options

Fig. 2.4. Comparison of probability densities for different maturity dates of DAX index options on December 1st, 2003 estimated for each single maturity date separately ('SMOOTHNESS') and calculated by backward induction using the Rubinstein (1994) algorithm ('RUBINSTEIN').

To reveal the differences between our algorithm and the Jackwerth and Rubinstein (1996) algorithm take a look at Fig. 2.4. We used the

March 2004 IRNPD recovered by the method from Jackwerth and Rubinstein (1996) to calculate the complete implied binomial tree with Rubinstein's (1994) backward induction procedure and then calculated the IRNPD for the earlier maturity dates from this implied binomial tree. Obviously, the IRNPDs of earlier maturity dates converge to the benchmark distribution if we work backwards using the Rubinstein (1994) algorithm. This is due to the assumption of equal path probabilities. Moreover, this is the reason why implied binomial trees for two different maturity dates constructed by Rubinstein's method are inconsistent for the overlapping period, i.e. between T_0 and the shorter maturity date. For example, compare the IRNPDs for December 2003 DAX index options in Fig. 2.4.a obtained by separate estimation using Jackwerth and Rubinstein (1996) ('SMOOTHNESS') and recovered from backward induction ('RUBINSTEIN'). The latter is similar to the benchmark distribution and therefore cannot fit the volatility smile for this maturity date, since the benchmark distribution is a log-normal distribution with constant volatility parameter.

If we use the standard binomial tree method for the state-space specification, we get many states with zero probability, which is not desirable. Therefore, we examine another method to specify the state space at the different maturity dates. In the previous example, we used the same up- and down-parameter to specify the state space. Now we use a confidence interval to specify the state space for each maturity date T_l.[4] Hence, we estimate the implied volatility $\hat{\sigma}_{T_l}$, which minimizes the sum of the squared pricing errors for each maturity date. Then we take this volatility parameter to calculate the following confidence interval as:

$$CI_l = [S_{T_0}g(-10), S_{T_0}g(5)] \qquad (2.13)$$

with

$$g(x) = \exp\left((r - \frac{\hat{\sigma}_{T_l}^2}{2})T_l + x\hat{\sigma}_{T_l}\sqrt{T_l}\right) \qquad (2.14)$$

Therefore, the value of the underlying at node (T_l, j) is given by

$$S_{T_l,j} = S_{T_0}\exp\left((r - \frac{\hat{\sigma}_{T_l}^2}{2})T_l + f(j)\hat{\sigma}_{T_l}\sqrt{T_l}\right), \qquad j = 1, ..., N_l \quad (2.15)$$

with

---

[4] A similar approach was proposed by Andersen and Brotherton-Ratcliffe (1998) to ensure that only statistically significant underlying prices are captured by the state-space specification.

$$f(j) = \frac{15.0(j-1) - 10.0(N_l - 1)}{N_l - 1} \qquad (2.16)$$

Presuming that $(r - \frac{1}{2}\hat{\sigma}_{T_l}^2)$ and $\hat{\sigma}_{T_l}$ are the correct moments of the underlying distribution, this leads to:

$$\mathbb{P}\left(S_{T_l} \in CI_l\right) > 0.99999. \qquad (2.17)$$

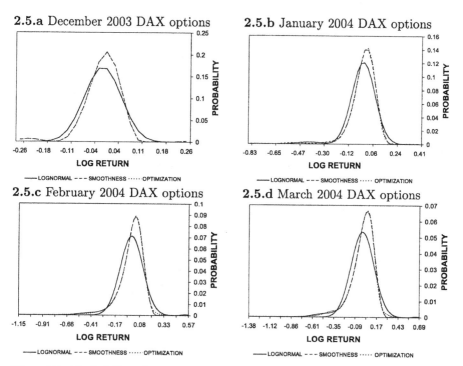

2.5.a December 2003 DAX options

2.5.b January 2004 DAX options

2.5.c February 2004 DAX options

2.5.d March 2004 DAX options

Fig. 2.5. Probability densities for different maturity dates of DAX index options on December 1st, 2003 with state-space specification using confidence interval.

This state-space specification is used to recover the IRNPD in Fig. 2.5. In particular, in Fig. 2.5.c and 2.5.d one can see that the Jackwerth and Rubinstein (1996) algorithm produces fat tails on the left-hand side of the distribution. In the Brown and Toft (1999) algorithm, this can lead to problems as described in Sect. 2.3.3. The main differences can be found in the transition probabilities for each sub-tree. The transition probability distributions in the standard state space setting are remarkably smooth for the sub-trees emanating from high and low node indices. This produces inconsistent values for the standard deviation of these transition probability distributions. Therefore, we conclude that the traded benchmark options contain only little information about the implied distribution far away from the current stock price and, therefore, that the benchmark options have no significant influence on the shape of the implied distribution. Using a confidence interval as given in (2.13) to specify the state space mitigates this problem. Table 2.1 reports the annualized mean and standard deviation of the implied transition probability distribution for sub-trees emanating from different node indices at January 2004. Using the standard method to specify the state space, we find that the standard deviations increase significantly as the market moves up. This contradicts the typical finding of a negative correlation between volatility and asset price. Therefore, we recommend specifying the state space in such a way that only statistically significant underlying prices are captured.

Table 2.1. Annualized Mean and Standard Deviation of Implied Risk-Neutral Probability Distributions and Implied Transition Probability Distributions

The first two rows report the annualized mean and standard deviation of the IRNPDs for January 2004 and February 2004. The third through thirteenth rows report the annualized mean and standard deviation of the implied transition probability distributions for subtrees emanating from different node indices at January 2004. The asset prices in the second and sixth column indicate the starting value of the corresponding subtree at January 2004. The index level at T_0 is 3799.505.

Standard Method			Confidence Interval		
Maturity	Mean	Std.Dev.	Maturity	Mean	Std.Dev.
$T_{2(\text{Jan }04)}$	-0.0091	0.2533	$T_{2(\text{Jan }04)}$	-0.0107	0.2616
$T_{3(\text{Feb }04)}$	-0.0123	0.2683	$T_{3(\text{Feb }04)}$	-0.0121	0.2681

Node at T_2	Asset Price	Mean	Std. Dev.	Node at T_2	Asset Price	Mean	Std. Dev.
50	7030.74	-0.1907	0.6507	50	5748.65	-0.0148	0.2710
45	6200.91	-0.1914	0.6519	45	5282.80	-0.0224	0.2980
40	5469.02	-0.1841	0.6404	40	4854.70	-0.0308	0.3254
35	4823.52	-0.1239	0.5382	35	4461.29	-0.0215	0.2936
30	4254.20	0.0008	0.2039	30	4099.76	0.0043	0.1861
25	3752.08	-0.0061	0.2358	25	3767.53	-0.0034	0.2238
20	3309.23	-0.0726	0.4367	20	3462.22	-0.0579	0.3987
15	2918.64	-0.1696	0.6191	15	3181.66	-0.0818	0.4504
10	2574.16	-0.1911	0.6511	10	2923.83	-0.0676	0.4161
5	2270.33	-0.1899	0.6495	5	2686.89	-0.0223	0.2946
1	2053.30	-0.1906	0.6506	1	2511.24	-0.0138	0.2650

We now use the same data as before to construct an implied multinomial tree to demonstrate the flexibility of our algorithm. So, we first have to estimate a volatility function. Alternatively, one can also use an interpolation method to calculate missing values. However, we prefer to approximate the volatility function, which leads to good results as shown by Dumas et al. (1998). We choose orthogonal Laguerre polynomials up to order eight in time to maturity, the moneyness, and their cross-product to fit the volatility function. Under this specification is the root mean squared error 0.0023 and the R^2 is equal to 0.9968. The resulting volatility surface is depicted in Fig. 2.6.

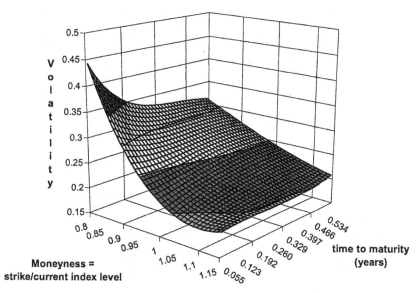

Fig. 2.6. Fitted volatility surface of DAX index options using orthogonal Laguerre polynomials.

This volatility function is now used to construct an implied multinomial tree with five branches, i.e. we set $M_t \equiv M = 5$. Numerical tests indicate that trinomial trees do not have enough degrees of freedom.[5] For the state-space specification, we use the volatility, which minimizes the sum of squared pricing errors over all maturity dates. Then we define $u = \hat{\sigma}\sqrt{4.0\frac{\Delta t}{M-1}}$ for all time steps. The value of the underlying at node (t, j) is given by

---

[5] Trinomial trees fit the prices of traded benchmark options within the bid-ask spread. However, in-the-money options are all undervalued, since it is difficult to generate fat tails on the left-hand side of the implied probability distribution.

$$S_{t,j} = S_0 \exp\left(u(j - \frac{(M-1)(t+1)}{2})\right) \qquad (2.18)$$

, where $t = 1, ..., T_n - T_0$ and $j = 1, ..., (M-1)(t+1)$. Since we do not have benchmark options for every time step, we construct a synthetic set. For each day $t = 1, ..., T_n - T_0$, we use the strike prices of those traded benchmark options which mature next and the corresponding values of the volatility function to calculate prices for European call options. Whenever $t = T_l - T_0$ for $l = 1, ..., n$, we use the market prices of the benchmark options of that maturity date to run the algorithm. Furthermore, we set $\alpha = 10^{-8}$, since the transition probabilities in each sub-tree will mainly fulfil the no-arbitrage conditions. All other parameters remain unchanged. In Fig. 2.7 we show the resulting IRNPD using our algorithm, as well as the IRNPD recovered for each single maturity date by using the Jackwerth–Rubinstein procedure. As the benchmark, we also plot the log-normal distribution.

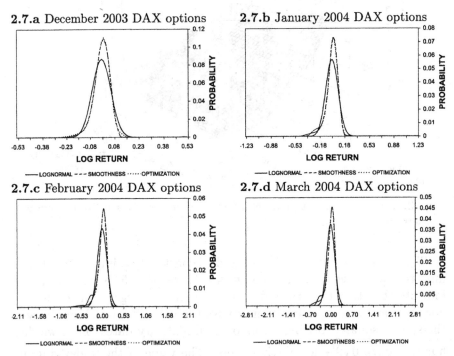

Fig. 2.7. Probability densities for different maturity dates of DAX index options on December 1st, 2003 using an implied multinomial tree with five branches.

We come remarkably close to the IRNPD inferred from the market prices by the method of Jackwerth and Rubinstein (1996) for each single maturity date separately. However, note that we get the IRNPDs by a forward algorithm that reproduces the market prices of all maturity dates correctly and not just for a single maturity date. The implied multinomial trees have some advantages to the implied binomial tree constructed with our method. First, the implied multinomial trees are recombining and not only semi-recombining. Furthermore, we can generate the IRNPD for each day. This simplifies the valuation of new options, since we work with a unique state space and plain-vanilla European-style options that can directly be evaluated by using their respective IRNPD. Additionally, multinomial trees with an odd number of branches are more appropriate to price Barrier options.[6] For this purpose, one should specify the state space in such a way that the barrier is hit exactly. This is not a problem, since the specification of the state space is arbitrary.

There is also another interesting point to emphasize when using multinomial trees. In binomial tree models, the market is dynamically complete, since we have two states in each sub-tree and two given basis assets to duplicate a contingent claim. This holds no longer for multinomial trees. Buraschi and Jackwerth (2001) show that for spanning purposes the returns of in-the-money and out-of-the-money options are needed. The multinomial tree can also include further options as basis assets such that the market is again dynamically complete. For our example, we propose to include an in-the-money, at-the-money, and out-of-the money option, respectively, besides the underlying asset and money-market account. Then we are able to construct hedge-portfolios of these five assets to price exotic options by no-arbitrage using standard replication arguments.

Finally, we want to point out that the multinomial setting can also be used to calibrate models with more than one state variable. Kamrad and Ritchken (1991) propose a method to approximate multinomial models with k state variables. The special case of two state variables is similar to our example for a multinomial tree with five branches. To construct an implied tree we have to add further constraints for the

---

[6] This was pointed out by Ritchken (1995) and Cheuk and Vorst (1996) for the special case of trinomial trees. Boyle and Lau (1994) have shown that using a naive application of binomial trees to price barrier options leads to significant pricing errors for such options.

second state variable to the optimization problem. This means that we get a forward constraint for the second state variable similar to (2.5).[7]

2.5 Conclusion

In this chapter, we have developed a new, more flexible and powerful method to construct risk-neutral, arbitrage-free, semi-recombining implied binomial trees that are consistent with given market prices of liquid-traded options. The advantage of our method for constructing implied binomial trees is that no interpolation or extrapolation steps are necessary and no prior guess about the benchmark distribution is required. This is achieved by using a 'smoothness criterion' to recover the implied risk-neutral probability distribution. Additionally, we have to solve a quadratic programming optimization problem with linear inequality constraints, which can be easily solved with standard software. Furthermore, our method uses all the available information on market prices to estimate the IRNPD, since the IRNPD of each maturity date incorporates the IRNPDs of all previous maturity dates. Under the additional assumption that a volatility function exists, the method can be used to construct arbitrage-free, risk-neutral, recombining implied multinomial trees. As a result, we are able to price and hedge many plain-vanilla and exotic options in accordance with given market prices.

Further research should examine the empirical performance of the method and compare it to existing approaches in a more extensive test. Here, it is of special interest which method performs better – constructing implied binomial trees or constructing implied multinomial trees. This is equivalent to the question of whether the assumption of equal path probabilities in each sub-tree or the assumption of the existence of a volatility function leads to better empirical results.

2.6 Appendix

Reconsider the optimization problem:

$$
\begin{aligned}
\min_{q_{j,m}} \alpha \sum_{j=1}^{N_l} \sum_{m=1}^{M_l} &\left(q_{j,m-1} - 2q_{j,m} + q_{j,m+1}\right)^2 \\
&+ \beta \sum_{k=1}^{N_{l+1}} \left(Q_{T_{l+1},k-1} - 2Q_{T_{l+1},k} + Q_{T_{l+1},k+1}\right)^2 \\
&+ \gamma \sum_{i=1}^{O} \frac{1}{w_i} \left(\sum_{k=1}^{N_{l+1}} Q_{T_{l+1},k} CF_{k,i} - CM_i \exp\left(rN_{l+1}\Delta t\right)\right)^2
\end{aligned}
$$

---

[7] This is only the case when the second state variable represent a price process.

subject to

$$\sum_{m=1}^{M_l} q_{j,m} = 1 \quad \text{for } j = 1, ..., N_l,$$

$$q_{j,m} \geq 0 \quad \text{for } j = 1, ..., N_l, \ m = 1, ..., M_l,$$

$$\sum_{m=1}^{M_l} q_{j,m} S_{T_{l+1},j+m-1} = S_{T_l,j} \exp\left(r\Delta_l \Delta t\right) \quad \text{for } j = 1, ..., N_l,$$

and define:

$$\hat{\mathbf{Q}} = \left(\begin{pmatrix} Q_{T_l,N_l}\mathbf{E}_{M_l} \\ \mathbf{0}_{(1\times M_l)} \\ \vdots \\ \vdots \\ \mathbf{0}_{(1\times M_l)} \end{pmatrix} \cdots \begin{pmatrix} \mathbf{0}_{(1\times M_l)} \\ \vdots \\ \mathbf{0}_{(1\times M_l)} \\ Q_{T_l,2}\mathbf{E}_{M_l} \\ \mathbf{0}_{(1\times M_l)} \end{pmatrix} \begin{pmatrix} \mathbf{0}_{(1\times M_l)} \\ \vdots \\ \vdots \\ \mathbf{0}_{(1\times M_l)} \\ Q_{T_l,1}\mathbf{E}_{M_l} \end{pmatrix} \right)_{(N_{l+1}\times N_l M_l)}$$

$$\mathbf{q} = \begin{pmatrix} q_{N_l,M_l} \\ \vdots \\ q_{N_l,1} \\ \vdots \\ q_{1,M_l} \\ \vdots \\ q_{1,1} \end{pmatrix}_{(N_l M_l \times 1)} \qquad \mathbf{B}_n = \begin{pmatrix} -2 & 1 & 0 & \cdots & \cdots & 0 \\ 1 & -2 & 1 & \ddots & & \vdots \\ 0 & 1 & -2 & 1 & \ddots & \vdots \\ \vdots & \ddots & \ddots & \ddots & \ddots & 0 \\ \vdots & & \ddots & 1 & -2 & 1 \\ 0 & \cdots & \cdots & 0 & 1 & -2 \end{pmatrix}_{(n\times n)}$$

$$\hat{\mathbf{B}} = \begin{pmatrix} \mathbf{B}_{M_l} & & \bigcirc \\ & \ddots & \\ \bigcirc & & \mathbf{B}_{M_l} \end{pmatrix}_{(N_l M_l \times N_l M_l)} \qquad \mathbf{W} = \begin{pmatrix} \frac{1}{w_1} & & \bigcirc \\ & \ddots & \\ \bigcirc & & \frac{1}{w_O} \end{pmatrix}_{(O\times O)}$$

$$\mathbf{C}_F = \begin{pmatrix} CF_{1,N_{l+1}} & \cdots & CF_{1,1} \\ \vdots & & \vdots \\ CF_{O,N_{l+1}} & \cdots & CF_{O,1} \end{pmatrix}_{(O\times N_{l+1})} \qquad \mathbf{s}_l = \begin{pmatrix} S_{T_l,N_l} \\ \vdots \\ S_{T_l,1} \end{pmatrix}_{(N_l\times 1)}$$

$$\mathbf{C}_M = \exp\left(rT_{l+1}\right) \begin{pmatrix} CM_1 \\ \vdots \\ CM_O \end{pmatrix}_{(O\times 1)} \qquad \mathbf{1}_n = \begin{pmatrix} 1 \\ \vdots \\ 1 \end{pmatrix}_{(n\times 1)} \qquad \mathbf{0}_n = \begin{pmatrix} 0 \\ \vdots \\ 0 \end{pmatrix}_{(n\times 1)}$$

$$\mathbf{A} = \begin{pmatrix} 1 & \cdots & 1 & 0 & \cdots & \cdots & 0 \\ \vdots & & & & & & \vdots \\ 0 & \cdots & \cdots & 0 & 1 & \cdots & 1 \\ S_{T_{l+1},N_{l+1}} & \cdots & S_{T_{l+1},N_{l+1}-M_l+1} & 0 & \cdots & \cdots & 0 \\ \vdots & & & & & & \vdots \\ 0 & \cdots & & 0 & S_{T_{l+1},M_l} & \cdots & S_{T_{l+1},1} \end{pmatrix}_{(2N_l \times N_l M_l)}$$

where $\mathbf{E}_{M_l}$ is the $M_l \times M_l$ identity matrix and n are an arbitrarily chosen number to define the size of the vector or matrix, respectively.

Given this definitions we can calculate the IRNPD at time T_{l+1} as follows:

$$\mathbf{Q}_{l+1} = \begin{pmatrix} Q_{T_{l+1},N_{l+1}} \\ \vdots \\ Q_{T_{l+1},1} \end{pmatrix}_{(N_{l+1}\times 1)} = \hat{\mathbf{Q}}\mathbf{q}$$

Then we are able to rewrite the optimization problem as:

$$\min_{\mathbf{q}} \ \alpha \mathbf{q}'\hat{\mathbf{B}}'\hat{\mathbf{B}}\mathbf{q}$$
$$+ \beta \mathbf{q}'\hat{\mathbf{Q}}'\mathbf{B}'_{N_{l+1}}\mathbf{B}_{N_{l+1}}\hat{\mathbf{Q}}\mathbf{q}$$
$$+ \gamma (\mathbf{C}_F\hat{\mathbf{Q}}\mathbf{q} - \mathbf{C}_M)'\mathbf{W}(\mathbf{C}_F\hat{\mathbf{Q}}\mathbf{q} - \mathbf{C}_M)$$

subject to

$$\mathbf{q} \geq \mathbf{0}_{(N_l M_l \times 1)}$$

$$\mathbf{A}\mathbf{q} = \mathbf{y} = \begin{pmatrix} \mathbf{1}_{N_l} \\ \mathbf{s}_l \end{pmatrix}_{(2N_l \times 1)}$$

After some rearrangements and redefinitions the objective function reduces to:

$$\min_{\mathbf{q}} \ \ \mathbf{q}'\mathbf{H}\mathbf{q} + \mathbf{b}'\mathbf{q} + c$$

where

$$\mathbf{H} := \alpha \hat{\mathbf{B}}'\hat{\mathbf{B}} + \hat{\mathbf{Q}}'(\beta \mathbf{B}'_{N_{l+1}}\mathbf{B}_{N_{l+1}} + \gamma \mathbf{C}'_F\mathbf{W}\mathbf{C}_F)\hat{\mathbf{Q}}$$
$$\mathbf{b} := -2\gamma(\mathbf{C}_F\hat{\mathbf{Q}})'\mathbf{W}\mathbf{C}_M$$
$$c := \gamma \mathbf{C}'_M\mathbf{W}\mathbf{C}_M$$

References

Aït-Sahalia, Y. and A.W. Lo (1998): Nonparametric Estimation of State-Price Densities Implicit in Financial Asset Prices, *The Journal of Finance*, 53(2):499–547.

Andersen, L.B.G. and R. Brotherton-Ratcliffe (1998): The Equity Option Volatility Smile: An Implicit Finite-Difference Approach, *The Journal of Computational Finance*, 1(2):5–37.

Bakshi, G., C. Cao, and Z. Chen (1997): Empirical Performance of Alternative Option Pricing Models, *The Journal of Finance*, 52(5):2003–2049.

Barle, S. and N. Cakici (1998): How to Grow a Smiling Tree, *Journal of Financial Engineering*, 7(2):127–146.

Black, F. and M. Scholes (1973): The Valuation of Options and Corporate Liabilities, *Journal of Political Economy*, 81(3):637–654.

Boyle, P. and S.H. Lau (1994): Bumping Up Against the Barrier with the Binomial Method, *The Journal of Derivatives*, 1(4):6–14.

Brown, G. and K.B. Toft (1999): Constructing Binomial Trees from Multiple Implied Probability Distributions, *The Journal of Derivatives*, 7(2):83–100.

Buraschi, A. and J.C. Jackwerth (2001): The Price of a Smile: Hedging and Spanning in Option Markets, *The Review of Financial Studies*, 14(2):495–527.

Cheuk, T.H.F. and T.C.F. Vorst (1996): Complex Barrier Options, *The Journal of Derivatives*, 4(1):8–22.

Cox, J.C., S.A. Ross, and M. Rubinstein (1979): Option Pricing: A Simplified Approach, *Journal of Financial Economics*, 7:229–263.

Derman, E. and I. Kani (1994): Riding on a Smile, *RISK*, 7(2):32–39.

Derman, E., I. Kani, and N. Chriss (1996): Implied Trinomial Trees of the Volatility Smile, *The Journal of Derivatives*, 3(4):7–22.

Dumas, B., J. Fleming, and R.E. Whaley (1998): Implied Volatility Functions: Empirical Tests, *The Journal of Finance*, 53(6):2059–2106.

Dupire, B. (1994): Pricing with a Smile, *RISK*, 7(1):18–20.

Jackwerth, J.C. (1997): Generalized Binomial Trees, *The Journal of Derivatives*, 5(2):7–17.

Jackwerth, J.C. (1999): Option Implied Risk-Neutral Distributions and Implied Binomial Trees: A Literature Review, *The Journal of Derivatives*, 7(2):66–81.

Jackwerth, J.C. and M. Rubinstein (1996): Recovering Probability Distributions from Option Prices, *The Journal of Finance*, 51(5):1611–1631.

Kamrad, B. and P. Ritchken (1991): Multinomial Approximating Models for Options with K State Variables, *Management Science*, 37(12):1640–1652.

Ritchken, P. (1995): On Pricing Barrier Options, *The Journal of Derivatives*, 3(2):19–28.

Rubinstein, M. (1994): Implied Binomial Trees, *The Journal of Finance*, 49(3):771–818.

Skiadopoulos, G. (2001): Volatility Smile Consistent Option Models: A Survey, *International Journal of Theoretical and Applied Finance*, 4(3):403–437.

3

Market-Conform Option Valuation: An Empirical Assessment of Alternative Approaches

3.1 Introduction

The classical Black–Scholes model assumes that the price of the underlying asset follows a geometric Brownian motion with constant volatility, which is the only unobservable parameter in the model. Consequently, all options on the same asset should produce the same implied volatility. However, many empirical studies, e.g. Rubinstein (1994), Bakshi et al. (1997), Dumas et al. (1998), Aït-Sahalia and Lo (1998), and Tompkins (2001a,b), document the existence of a systematic relationship between the implied volatility and the strike prices or moneyness ratios of traded options (known as the *volatility smile*). On the other hand, the *term structure of implied volatility* describes the relation of contemporaneous implied volatilities of options on the same underlying asset and with the same strike prices but with *different times to maturity*. Such volatility patterns contradict the assumption of a geometric Brownian motion, which would imply a constant or, at most, deterministically time-varying volatility. Similarly, Jackwerth and Rubinstein (1996) show that the implied risk-neutral probability densities are heavily skewed to the left and are highly leptokurtic, which contradicts the log-normality assumption in the Black and Scholes (1973) model.

In the light of these shortcomings, a large number of deterministic and stochastic volatility models were proposed to relax the restrictive assumptions in the Black–Scholes model. These models are designed to incorporate the implied volatility structure. The objective of this chapter is to perform an empirical test of different models to price a common set of options in accordance with given market prices. We are especially interested in the relative performance of the competing models. In our analysis, we categorize the different models into three classes: (i) parametric option pricing models; (ii) deterministic volatility models; and (iii) non-parametric pricing models, which rely solely on observed option prices without specifying a stochastic process a priori.

In the first class we examine the 'classical' Black–Scholes model (BS) and the stochastic volatility model (H) from Heston (1993). For the deterministic volatility models, we specify a volatility function (DVF), as proposed by Dumas et al. (1998), as well as a 'naive-trader-rule' (NTR) to incorporate the volatility smile into the Black–Scholes model. For the non-parametric models, we check the implied tree model (IBT) from Herwig (2005) and the Weighted Monte Carlo approach from Avellaneda et al. (2001), where the Black–Scholes model (WBS) and the Heston model (WH) are used to generate the sample paths. We use the Black–Scholes model as a benchmark that we presume exhibits the worst pricing performance, since it explicitly assumes a constant volatility of the underlying.

We implement each model by estimating the implied volatility and other structural parameters from observed option prices at the EUREX for each trading day. This approach is common in the existing literature (e.g. Bakshi et al. (1997) or Belledin and Schlag (1999)). Estimation of the Black–Scholes model volatility for a daily frequency is somehow ad-hoc, since it is basically incompatible with the assumption of a constant volatility over time. An analogous argument also holds for various parameters in the other models. However, this approach is widely used in practice, and we follow this convention to ensure an equal chance for each model.

It is not always possible to replicate observable market prices of liquid instruments with the existing parametric option pricing models. This is due to the fact that often more instruments are traded than there are model parameters. Therefore, a correct specification of parametric option pricing models consistent with the prices of all traded options is, in most cases, impossible in a real-world application.

In contrast to other studies, our analysis focuses on an out-of-sample pricing performance comparison instead of testing the out-of-sample predictive power for future implied volatilities of the different models. The latter kind of test is performed, e.g. by Bakshi et al. (1997), Dumas et al. (1998), and Jackwerth and Rubinstein (2001). For our study, we use quotes from two competing derivative markets to examine the pricing performance of the models. The key idea for this procedure is to determine the price for a newly issued option. That is why we assume that at one market we observe the correct prices of such newly issued options, whereas the other market is assumed to provide the basis with all previously issued options. Therefore, we take market quotes for DAX index options from the EUREX, the world's largest futures and options exchange, and use these quotes to calibrate the pricing models. Afterwards, we use the calibrated models to price different types of options traded at the European Warrant Exchange (EUWAX), the world's largest derivative exchange in terms of listed bank-issued instruments, and compare the model prices to the market quotes.

In Sect. 3.2, we describe each model briefly with respect to its assumptions and the associated valuation procedure, as well as the techniques needed for calibration. In Sect. 3.3, we introduce our dataset. Then we conduct the empirical tests in Sect. 3.4. In particular, Sect. 3.4.1 and Sect. 3.4.2 deal with inferring American call option prices and European knock-out option prices on a non-dividend paying underlying from European call option prices traded at the EUREX. We conclude the chapter by summarizing the main results in Sect. 3.5.

3.2 Alternative Option Pricing Models

3.2.1 Parametric Option Pricing Models

The simplest parametric option pricing model was proposed by Black and Scholes (1973), who assume that the underlying follows a geometric Brownian motion with constant volatility. The dynamics of the underlying under the risk-neutralized measure are then given by

$$dS_t = rS_tdt + \sigma S_t dW_t, \tag{3.1}$$

where r denotes the risk-free interest rate, σ is the constant volatility, and dW_t denotes the increment of a standard Wiener process. To estimate the volatility parameter σ we minimize the sum of squared pricing errors

$$\min_{\sigma} \sum_{i=1}^{B} \left[C_i^M - C_i^{BS}(\sigma) \right]^2, \tag{3.2}$$

where C_i^M is the mid market price of option i, B is the number of benchmark options quoted at the EUREX, and $C_i^{BS}(\sigma)$ denotes the price of option i calculated with the Black–Scholes formula. We repeat this calibration for each trading date. Note that the resulting volatility will vary from day to day.

A more sophisticated parametric option pricing model was proposed by Heston (1993). This model incorporates stochastic variance in addition to the stochastic process of the underlying. Thereby, the volatility is formulated in terms of the instantaneous variance and modelled as a mean-reverting square-root process. The dynamics of the underlying stock and its instantaneous variance are given under the risk-neutralized measure by

$$dS_t = rS_t dt + \sigma_t S_t dW_t^{(1)},$$
(3.3)

$$d\sigma_t^2 = \kappa(\theta - \sigma_t^2)dt + \eta\sigma_t dW_t^{(2)},$$
(3.4)

where the increments of the Wiener processes, $dW_t^{(1)}$ and $dW_t^{(2)}$ are correlated with constant correlation coefficient ρ. The parameters κ, θ, and η represent the speed of adjustment, the long-run mean, and the volatility of the instantaneous variance σ_t^2. Heston (1993) has shown that a closed-form pricing formula for standard European options can be derived by using Fourier inversion techniques. Therefore, it is convenient to estimate the unknown parameters κ, θ, η, ρ and the initial value of the instantaneous variance $\sigma_{t_0}^2$ for the given benchmark instruments in the sense of least-squares by

$$\min_{\sigma_{t_0}^2,\kappa,\theta,\eta,\rho} \sum_{i=1}^{B} \left[C_i^M - C_i^H(\sigma_{t_0}^2,\kappa,\theta,\eta,\rho) \right]^2,$$
(3.5)

where $C_i^H(\sigma_{t_0}^2,\kappa,\theta,\eta,\rho)$ is determined by the pricing formula of the Heston model for standard European options.

More complex parametric models incorporate stochastic interest rates and/or stochastic jumps in addition to the stochastic volatility. However, Bakshi et al. (1997) have shown that adding stochastic jumps or stochastic interest rates as further market risk factors only slightly improves the pricing and hedging performance. Therefore, we focus our analysis on the stochastic volatility model, since more sophisticated parametric models complicate the implementation and estimation of parameters without yielding any distinct performance improvement.

The main advantage of the parametric option pricing models is that the closed-form pricing formulas are available for European plain-vanilla options. Moreover, in the Black–Scholes environment there are also closed-form solutions for certain types of exotic derivatives, e.g. Barrier options, Asian options, or Digital options. However, the existence of these closed-form pricing formulas is a result of restrictive assumptions in the Black–Scholes model. For more complex models, in general closed-form solutions do not exist for exotic options. Numerical methods are often needed to solve those pricing problems. Derivatives with early-exercise features are the most popular options for which no closed-form solutions exist, e.g. American put options. To solve the pricing problem, the underlying model, the model parameters, and an adequate numerical method are needed.

3.2.2 Deterministic Volatility Models

The main characteristic of deterministic volatility models is that they parameterize the observed implied volatility structure and combine it with the Black–Scholes model. This can be done either by estimating the implied volatility surface as a function of time to maturity τ and the strike price K or by a simple interpolation or extrapolation of the given implied volatilities. The latter method is often used by traders to accommodate observed smile patterns and is usually called 'naive-trader-rule'.

Dumas et al. (1998) test different specifications for a volatility function depending on time to maturity τ and strike price K. In accordance with Dumas et al. (1998), we use the following volatility function (DVF)

$$\sigma(K,\tau) = \max(0.01, \alpha_0 + \alpha_1 K + \alpha_2 K^2 + \alpha_3 \tau + \alpha_4 K\tau), \qquad (3.6)$$

whereby a minimum value for the volatility is imposed to prevent negative values. We estimate the unknown parameters α_j, $j = 0, ..., 4$ at each trading day for the given benchmark instruments by solving the following optimization problem:

$$\min_{\substack{\alpha_j \\ j=0,...,4}} \sum_{i=1}^{B} \left[C_i^M - C_i^{BS}(\sigma(K,\tau)) \right]^2. \qquad (3.7)$$

In contrast to Dumas et al. (1998), who minimize the sum of squared pricing errors between the market prices and the price given by the solution of the forward partial differential equation to estimate the unknown parameters, our procedure to estimate the volatility function is a simplified version. Our applied estimation procedure coincides with the ad hoc model specification used by Dumas et al. (1998). This ad hoc model has uniformly smaller valuation errors than those of the more sophisticated solution of the forward partial equation. Due to this, we use this simplified version in our analysis. Consequently, we use the Black–Scholes formula in combination with the volatility function to find the theoretical prices for the benchmark options. While the Black–Scholes formula is generally used to extract implied volatilities from reported option quotes, it is also possible to revert the calculation to obtain option prices from estimated volatilities.

However, traders often use an even simpler method to account for observed smile patterns, the so-called naive-trader-rule (NTR). They do not use all observed option quotes to fit a volatility function but instead use an interpolation method of the nearest sampling points in the implied volatility grid. To simulate this NTR, we need to interpolate the implied volatilities of given option prices in strike and time dimensions. We use piecewise-linear interpolation, since the results obtained from more sophisticated methods, such as cubic splines or polynomial interpolation, are very similar. Note that in the NTR no more than four adjoining sampling points are used to determine the volatility, whereas the DVF approach uses all benchmark options to fit a functional form of the volatility curve.

We apply a two-step procedure to interpolate unknown volatilities by interpolating first the smile and then the term structure. Suppose we seek to determine a volatility for an option that matures between two maturity dates of traded options with a strike price which does not appear in the market. Then we interpolate for each maturity date the volatility smile to get the volatility for the corresponding strike price of the option at these maturity dates. Afterwards, we interpolate the term structure of volatilities using the interpolated values for the strike price at the two maturity dates. We use this interpolation sequence, since there are more strike prices available for each maturity date than there are different maturity dates.[1] Note that in contrast to estimation approaches, the NTR fits observed options quotes perfectly, since they are used as sampling points for the interpolation method. Furthermore, although the NTR does not rely on a theoretical basis, it is widely used in practice, which makes its investigation worthwhile.

3.2.3 Non-Parametric Option Pricing Models

Implied Trees

The construction of implied trees goes back to contributions by Rubinstein (1994), Derman and Kani (1994), and Dupire (1994). The aim of the models is to infer information from the given market prices of options. This 'first generation' of implied trees has some disadvantages as transition probabilities can become negative or time-inconsistent.

---

[1] The described algorithm is also used for extrapolation with the restriction that extrapolated values smaller than 0.01 are replaced by 0.01 to avoid negative volatilities and to ensure a minimum level of volatility.

Therefore, extended versions were suggested by Jackwerth and Rubinstein (1996), Derman et al. (1996), Jackwerth (1997), Barle and Cakici (1998), and Brown and Toft (1999). A recent contribution concerning the construction of implied trees was provided by Herwig (2005). This approach allows us to construct both implied binomial trees and implied multinomial trees. We focus our analysis on the construction principle for implied binomial trees (IBT).

Herwig (2005) proposes a forward algorithm to construct arbitrage-free implied trees. For each maturity date T_l, the implied risk-neutral probability distribution is estimated solving an optimization problem, $l = 1, ..., n$, where n is the number of available maturity dates. At each maturity date T_l, the transition probability $q_{j,m}$, $j = 1, ..., N_l$, $m = 1, ..., M_l$ to reach node $k = j + m - 1$ at maturity date T_{l+1} emanating from node j at maturity date T_l has to be estimated for all nodes N_l. Thereby, M_l is the number of reachable nodes at maturity date T_{l+1}, starting from node j at maturity date T_l. For illustration purposes, the structure of the implied tree from maturity date T_l to maturity date T_{l+1} is depicted in Fig. 3.1. Since the 'smoothness criterion' defined by Jackwerth and Rubinstein (1996) is used to identify the transition probabilities $q_{j,m}$, the following optimization has to be solved for each available maturity date:

$$
\min_{q_{j,m}} \alpha \sum_{j=1}^{N_l} \sum_{m=1}^{M_l} \left(q_{j,m-1} - 2q_{j,m} + q_{j,m+1} \right)^2
$$
$$
+ \beta \sum_{k=1}^{N_{l+1}} \left(Q_{T_{l+1},k-1} - 2Q_{T_{l+1},k} + Q_{T_{l+1},k+1} \right)^2 \tag{3.8}
$$
$$
+ \gamma \sum_{i=1}^{B} \frac{1}{w_i} \left(C_i^M \exp\left(rT_{l+1} \right) - \sum_{k=1}^{N_{l+1}} Q_{T_{l+1},k} CF_{k,i} \right)^2
$$

subject to

$$
\sum_{m=1}^{M_l} q_{j,m} = 1, \quad \text{for } j = 1, ..., N_l, \tag{3.9}
$$

$$
q_{j,m} \geq 0, \quad \text{for } j = 1, ..., N_l, \; m = 1, ..., M_l, \tag{3.10}
$$

$$
\sum_{m=1}^{M_l} q_{j,m} S_{T_{l+1},j+m-1} = S_{T_l,j} \exp\left(r(N_{l+1} - N_l)\Delta t \right), \quad \text{for } j = 1, ..., N_l,
$$
$$
\tag{3.11}
$$

where $Q_{T_{l+1},k}$ defines the implied risk-neutral probability to reach node k at maturity date T_{l+1} from t_0, $CF_{k,i}$ denotes the cash-flow of option i in state k, and $S_{l,k}$ is the nodal value of the underlying in state k at maturity date T_l. The size of one time step is $\Delta t = \frac{1}{365}$, which corresponds to one day. The weighting factors are specified as $w_i = C_i^M$

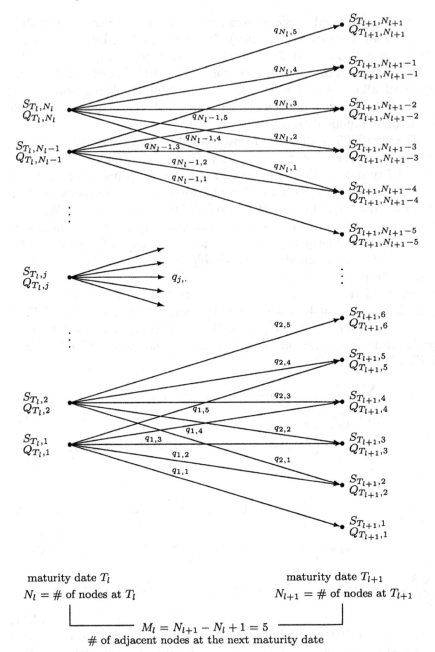

Fig. 3.1. Structure of an implied tree at the $l-th$ maturity date. For simplification M_l was set to five.

and for the penalty parameters we choose $\alpha = \frac{1}{N_\tau}$, $\beta = 1$, and $\gamma = \frac{1000}{N_{\tau+1}}$ as recommended in Herwig (2005).

Apart from these parameters, we must also specify the state space at each maturity date. For this purpose, we use a confidence interval for each maturity date T_l. This is done by estimating the implied volatility σ_l, which minimizes the sum of the squared pricing errors of traded options for each maturity date T_l. Then we take this volatility parameter to calculate the following confidence intervals (CI_l)

$$CI_l = [S_{t_0} g(-5), S_{t_0} g(5)] \tag{3.12}$$

with

$$g(x) = \exp\left((r - \frac{\hat{\sigma}_l^2}{2})T_l + x\hat{\sigma}_l\sqrt{T_l} \right) \tag{3.13}$$

Presuming that $(r - \frac{1}{2}\sigma_l^2)$ and σ_l^2 are the correct moments of the underlying distribution, this implies:

$$\mathbb{P}\left(S_{T_l} \in CI_l\right) > 0.99999. \tag{3.14}$$

Following this approach, we are able to construct semi-recombining, arbitrage-free implied binomial trees that are consistent with observed market prices. These implied trees can then be used to price a wide range of different option types.

Weighted Monte Carlo Approach

Avellaneda et al. (2001) propose an approach for reproducing spot prices of liquid and actively traded securities via Monte Carlo simulation. In contrast to the *classical* Monte Carlo approach, where all simulated paths have the same probability (or weight), Avellaneda et al. (2001) introduce different probabilities for each path in the simulation, i.e. the approach uses a weighted average over all sample paths to determine the option value instead of the simple mean. The aim is to determine the risk-neutral probabilities p_i for each path, which are determined such that the sum of weighted least-squares residuals (defined as the difference of the expected value of the discounted cash-flows of the benchmark instruments and the market prices of these benchmark instruments) is minimized, i.e.

$$\min_p \chi_w^2 = \frac{1}{2} \sum_{i=1}^{B} \frac{1}{w_j} (\mathbb{E}^p[CF_j] - C_j^M)^2, \tag{3.15}$$

where $\mathbb{E}^p[.]$ denotes the expectation operator under the risk-neutral measure p, χ_w^2 is the sum of weighted least-squares residuals, and w_j are positive weights. The relaxation of exact matching is justified by the existence of bid-ask-spreads and liquidity constraints and is recommended by Avellaneda et al. (2001) to avoid numerical problems in the optimization algorithm. Moreover, the risk-neutral probabilities p_i are as close as possible to the prior probabilities, i.e. they minimize the expression over all vectors q

$$D(q|u) = \log(M) + \sum_{i=1}^{M} q_i \log(q_i), \qquad (3.16)$$

where the distance measure $D(q|u)$ represents the relative entropy with uniform probabilities $1/M$ as priors and M as the number of generated paths.

To generate the sample paths in the Monte Carlo simulation we use the models by Black and Scholes (1973) and Heston (1993), i.e. we test the method by Avellaneda et al. (2001) with two different 'priors'. To estimate the model parameters, we use the procedure given in (3.2) and (3.5). We use the estimated parameters for the BS model (WBS) and the Heston model (WH) to generate the sample paths of (3.1) as well as of (3.3) and (3.4). Again, we choose one time step per day for the discretization of the processes, i.e. $\Delta t = \frac{1}{365}$ in the same manner as for the implied trees. Afterwards, we calibrate the simulated prices to the market prices to get the 'calibrated' paths probabilities p_i. We use $\frac{1}{w_j} = C_j \times 10^4$ as weighting parameters for the least-squares approximation in (3.15).

3.3 Data

We base our analysis on minute-by-minute quotes for DAX index options traded at EUREX. DAX index levels are taken from XETRA, while prices and quotes for American call options and European knock-out options are taken from EUWAX for the period from January 2, 2004 through June 30, 2004. The DAX index options traded at EUREX are standardized European-style and expire on the third Friday of the contract months, whereas options traded at EUWAX are bank-issued instruments, i.e. each issuer is free to choose any option characteristics. However, due to competition, the options are extensively harmonized.[2]

As already mentioned, we parameterize all pricing models to reproduce the observed option prices at EUREX. Afterwards, the models are used to price the traded options at EUWAX.

Since the DAX is a capital-weighted performance index, i.e. dividends are reinvested, we do not have to take care of the payout rate. We use the 3-month EURIBOR as proxy for the risk-free rate. Over the sample period, the median of this rate was 2.075 % p.a.

To obtain sets of option prices across a sufficient number of strike prices for the options traded at EUREX, we select the last option quotes in the five minute interval before 4.30 P.M. CET to calibrate the pricing models.[3] When the quote was not updated in this interval, we assume that no new information about the volatility structure has entered the market. We only use one single time point during the trading hours for each day in the sample to ease the computational burden. As a proxy for the market price of the options, we use the midpoint between bid and ask.

In order to reduce estimation uncertainty because of large bid-ask spreads or illiquid options, we apply four exclusion criteria to the EUREX options. First, we eliminate all options with less than 36 quotes on a given trading day to ensure a minimum update-frequency of the quotes. Second, as commonly done in the relevant literature, we eliminate options with a time to maturity of less than ten or more than 270 days. Third, we only use options with moneyness ratios, defined as the ratios of the current index level to the present value of the strike price, between 0.85 and 1.15. This is done to account for a lack of liquidity far away from the money, since deep in- and out-of-the-money options are not actively traded. Finally, we rule out all options with prices less than 10 euro cent, since, in general, the relative bid-ask spreads for such options are extremely large.

In a next step, general arbitrage violations have to be eliminated from the data. For this purpose, we first convert all put prices into call prices using the standard European put-call parity. Then we check for the following condition to hold:

---

[2] An extensive comparison of both markets, in particular focusing on market-microstructure and product differences, is performed by Bartram and Fehle (2004).

[3] We first convert the observed bid-ask quotes to the implied bid-ask volatilities with the underlying price at the respective time-point using the Black–Scholes formula. Afterwards, we reconvert this into the bid-ask quote at 4:30 P.M. CET with the observed underlying price at this time.

$$S_{t_0} \geq C_{t_0}(K,T) \geq (S_{t_0} - K \exp(-r(T - t_0)))^+. \qquad (3.17)$$

Afterwards, we check for violations of vertical and butterfly spreads to ensure the monotonicity and the convexity property of the call price function. All violating options are removed from the dataset and then reintroduced in such a way that all of the above tests are not violated and that the largest number of options can be applied.[4]

The final sample consists of 6,763 observations for the 127 trading days of the first half of 2004. The median time to maturity of the EUREX options in our sample is 0.211 years or 77 days, the median moneyness ratio is 1.005, and the median of the overall relative spread is 4.168%. The relative spread is given by the ratio between the absolute spread and the midpoint.

For the EUWAX-traded options the data selection process is simpler. Again, we focus on the last option quotes in the five minute interval before 4.30 P.M. CET and use the midpoint between bid and ask as a proxy for the market price of the options.

In our analysis, we only use options from the EUWAX with a time to maturity of less than or equal to the longest available time to maturity for EUREX options. The minimum time to maturity for all options is seven days. Furthermore, we eliminate a few obvious data errors, as well as a few quotes with excessive relative bid-ask spreads.

The American call option sample consists of 35,914 observations for the sample period. The median of time to maturity is 0.230 years or 84 days, the median moneyness ratio is 1.028, and the median of the overall relative spread is 0.783%.[5] The moneyness ratio ranges from 0.672 to 2.080.

The European knock-out option sample consists of 15,874 observations for the sample period, split into 9,026 down-and-out call options and 6,848 up-and-out put options. We only consider up-and-out put options and down-and-out call options where the barrier is equal to the strike price, since these are the typical barrier options traded at EUWAX. Due to this fact all knock-out options are in-the-money, i.e. the current index level is above (below) the strike price for the call (put) options.

We group the options into 15 categories according to their moneyness ratios and their time to maturity. According to the time to maturity, an option can be classified as (i) short-term (< 60 days); (ii)

---

[4] For the exact details of this procedure the interested reader is referred to Jackwerth and Rubinstein (1996).

[5] The difference between the relative spread at EUWAX and EUREX is also reported by Bartram and Fehle (2004), who argue that this is due to trading motives and brokerage costs.

medium term $(60 - 180$ days); and (iii) long term (> 180 days). The intervals for the five 'moneyness categories' are: $[0, 0.94)$ for deep-out-of-the-money options (DOTM), $[0.94, 0.98)$ for out-of-the-money options (OTM), $[0.98, 1.02)$ for at-the-money options (ATM), $[1.02, 1.06)$ for in-the-money (ITM) options, and $[1.06, \infty)$ for deep-in-the-money options (DITM).

Table 3.1. Descriptive Statistics of the EUREX Sample

In each cell, the entries are the median of the observed midpoint option prices, the median of the relative bid-ask spread (in parentheses, computed as the difference between ask and bid divided by the midpoint price), and the total number of observations (in brackets). The option quotes and the contemporaneous index levels were recorded between 4.25 P.M. CET and 4.30 P.M. CET daily for the period from January 2, 2004 to June 30, 2004. The moneyness ratio M is measured as the current index level divided by the present value of the strike price.

		Time to maturity			
		< 60 days	60 -180 days	> 180 days	All maturities
	$M < 0.94$	15.26 (0.111) [171]	41.25 (0.080) [687]	102.55 (0.042) [325]	46.40 (0.074) [1183]
M o n e y n e s s	$0.94 \leq M < 0.98$	33.98 (0.054) [615]	101.85 (0.035) [659]	206.83 (0.027) [161]	71.94 (0.040) [1435]
	$0.98 \leq M < 1.02$	96.50 (0.031) [638]	173.73 (0.028) [482]	269.27 (0.026) [106]	133.17 (0.030) [1226]
	$1.02 \leq M < 1.06$	188.34 (0.046) [460]	265.55 (0.032) [589]	345.12 (0.028) [166]	244.58 (0.035) [1215]
	$M \geq 1.06$	377.46 (0.097) [516]	419.94 (0.056) [921]	485.69 (0.033) [267]	423.73 (0.058) [1704]
	All moneyness categories	110.02 (0.049) [2400]	192.85 (0.042) [3338]	260.10 (0.031) [1025]	167.18 (0.042) [6763]

The characteristics of the samples with respect to price level, percentage spread, and number of observations for all examined option types are reported in Tables 3.1–3.4.[6] The observations for the EUREX

---

[6] EUWAX traded options exhibit an exchange ratio of 100 : 1, i.e. 100 options are needed to buy/sell one unit of the underlying. Therefore, we multiply all prices of the EUWAX traded options by 100 to get comparable figures.

Table 3.2. Descriptive Statistics of the EUWAX American Call Sample

In each cell, the entries are the median of the observed midpoint option prices multiplied by 100, the median of the relative bid-ask spread (in parentheses, computed as the difference between ask and bid divided by the midpoint price), and the total number of observations (in brackets). The option quotes and the contemporaneous index levels were recorded between 4.25 P.M. CET and 4.30 P.M. CET daily for the period from January 2, 2004 to June 30, 2004. The moneyness ratio M is measured as the current index level divided by the present value of the strike price.

		Time to maturity			
		< 60 days	60 -180 days	> 180 days	All maturities
M o n e y n e s s	$M < 0.94$	6.80 (0.149) [1876]	33.50 (0.036) [4061]	89.00 (0.017) [2022]	29.50 (0.040) [7959]
	$0.94 \leq M < 0.98$	37.00 (0.031) [1835]	116.00 (0.012) [2079]	224.50 (0.009) [749]	88.50 (0.014) [4663]
	$0.98 \leq M < 1.02$	103.00 (0.016) [1719]	196.00 (0.009) [1911]	315.75 (0.007) [662]	167.50 (0.011) [4292]
	$1.02 \leq M < 1.06$	206.00 (0.010) [1749]	297.00 (0.007) [1841]	398.50 (0.005) [663]	267.50 (0.008) [4253]
	$M \geq 1.06$	586.55 (0.004) [5432]	699.00 (0.003) [7229]	668.50 (0.003) [2086]	655.50 (0.003) [14747]
	All moneyness categories	206.00 (0.009) [12611]	277.00 (0.007) [17121]	313.50 (0.007) [6182]	259.00 (0.008) [35914]

sample (Table 3.1) are spread out almost evenly across the moneyness groups. The category containing long-term options has only 1,025 observations, compared to 2,400 short-term options and more than 3,000 mid-term options. The observed prices reveal a similar pattern with respect to their time to maturity. This pattern is also contained in the EUWAX American call option sample (Table 3.2). The number of observations for DOTM and DITM are slightly higher than for OTM, ATM, and ITM options, which is due to the 'open-ended' character of these subcategories.

For the EUWAX down-and-out call option sample (Table 3.3) and the EUWAX up-and-out put option sample (Table 3.4), we have an increasing number of observations from ATM to DITM options[7] and a

---

[7] Keep in mind that for put options the categories are reversed.

Table 3.3. Descriptive Statistics of the EUWAX Down-and-Out Call Sample

In each cell, the entries are the median of the observed midpoint option prices multiplied by 100, the median of the relative bid-ask spread (in parentheses, computed as the difference between ask and bid divided by the midpoint price), and the total number of observations (in brackets). The option quotes and the contemporaneous index levels were recorded between 4.25 P.M. CET and 4.30 P.M. CET daily for the period from January 2, 2004 to June 30, 2004. The moneyness ratio M is measured as the current index level divided by the present value of the strike price.

		Time to maturity			
		< 60 days	60 -180 days	> 180 days	All maturities
M		67.75	70.25	75.00	68.50
o	$0.98 \leq M < 1.02$	(0.027)	(0.028)	(0.027)	(0.027)
n		[408]	[204]	[1]	[613]
e		170.00	175.63	164.50	172.00
y	$1.02 \leq M < 1.06$	(0.011)	(0.011)	(0.012)	(0.011)
n		[1603]	[804]	[20]	[2427]
e		478.50	467.50	630.00	477.50
s	$M \geq 1.06$	(0.004)	(0.004)	(0.003)	(0.004)
s		[3867]	[1999]	[120]	[5986]
	All	337.50	336.00	571.00	340.00
	moneyness	(0.006)	(0.006)	(0.004)	(0.006)
	categories	[5878]	[3007]	[141]	[9026]

decreasing number of observations for longer times to maturity. This is due to the knock-out probability for these options. The observed prices reveal a similar pattern with respect to their moneyness, where the pattern with respect to their time to maturity is not as obvious as for the EUREX sample and the EUWAX American call option sample.

3.4 Empirical Results

To assess the performance of the different models three criteria are used which are defined as follows:

- The root mean squared euro valuation error (RMSE) is the square root of the average squared deviations of the model prices from the reported midpoint option prices

$$RMSE = \sqrt{\frac{1}{N} \sum_{i=1}^{N} (C_i^M - C_i^{\text{model}})^2}, \qquad (3.18)$$

where N is the number of options quoted at EUWAX.

Table 3.4. Descriptive Statistics of the EUWAX Up-and-Out Put Sample

In each cell, the entries are the median of the observed midpoint option prices multiplied by 100, the median of the relative bid-ask spread (in parentheses, computed as the difference between ask and bid divided by the midpoint price), and the total number of observations (in brackets). The option quotes and the contemporaneous index levels were recorded between 4.25 P.M. CET and 4.30 P.M. CET daily for the period from January 2, 2004 to June 30, 2004. The moneyness ratio M is measured as the current index level divided by the present value of the strike price.

		Time to maturity			
		< 60 days	60 -180 days	> 180 days	All maturities
M		395.50	394.25	690.25	401.50
o	$M < 0.94$	(0.005)	(0.005)	(0.003)	(0.005)
n		[2063]	[1106]	[164]	[3333]
e		169.00	164.00	185.50	168.00
y	$0.94 \leq M < 0.98$	(0.011)	(0.012)	(0.011)	(0.012)
n		[1720]	[979]	[45]	[2744]
e		65.50	62.00	61.50	64.50
s	$0.98 \leq M < 1.02$	(0.028)	(0.031)	(0.033)	(0.029)
s		[501]	[267]	[3]	[771]
	All	252.50	241.50	530.75	252.50
	moneyness	(0.008)	(0.008)	(0.004)	(0.008)
	categories	[4284]	[2352]	[212]	[6848]

- The mean outside error (MOE) is the average valuation error outside the bid-ask spread. The outside error is defined as the difference between the bid (ask) price and the model price, if the model price is below (above) the reported option bid (ask) price and is set equal to zero otherwise

$$MOE = \frac{1}{N} \sum_{i=1}^{N} e_i, \quad \text{where } e_i = \begin{cases} C_i^{\text{model}} - C_i^{\text{bid}} & \text{if } C_i^{\text{model}} < C_i^{\text{bid}} \\ C_i^{\text{model}} - C_i^{\text{ask}} & \text{if } C_i^{\text{model}} > C_i^{\text{ask}} \\ 0 & \text{otherwise} \end{cases}$$

(3.19)

where e_i is the outside valuation error, and C_i^{bid} (C_i^{ask}) is the observed bid (ask) price. Therefore a positive (negative) value of MOE means that the model prices are too high (low) on average.

- The median of the absolute percentage error (MAOPE) is the median absolute valuation error outside the bid-ask spread relative to the reported midpoint option prices ($\tilde{e}_i = |\frac{e_i}{C_i^M}|$). This measure illustrates the exactness with which each model fits within the quoted bid and ask prices.

These measures are used to compare the pricing performance of the different models in a real market environment.

Table 3.5 contains statistics on the in-sample performance of the different approaches. We only report the results for the complete sample, since in each subcategory the order for the different approaches are quite similar. As expected, the NTR has the best in-sample performance, since the observed quotes are used at sampling points, whereas the Black–Scholes model (BS) performs worst, since there is only one free parameter for calibration. The MAOPE of the non-parametric pricing models WBS, WH, and IBT are less than half the overall bid-ask spread, which indicates that the fitted model prices are mostly within the bid-ask spread for these models. Thereby, the Weighted Monte Carlo models are more accurate than the implied binomial tree model. DVF and H perform quite similar, with the difference that the MOE is lower for the Heston model.

Table 3.5. In-Sample-Fit Statistics of the EUREX Sample

RMSE is the root mean squared euro valuation error. MOE is the mean valuation error outside the observed bid/ask quotes (a positive value indicates that the theoretical value exceeds the ask price on average and a negative value indicates the theoretical value is below the bid price). MAOPE is the median absolute percentage valuation error outside the observed bid/ask quotes (in percent). The sample period is January 2, 2004 to June 30, 2004, with a total of 6,763 observations.

Model/Approach	RMSE	MOE	MAOPE
Black–Scholes (BS)	7.908	-0.082	2.88
Naive-trader-rule (NTR)	0.000	0.000	0.00
Volatility function (DVF)	5.517	0.117	1.74
Heston (H)	5.736	0.015	1.75
Weighted Monte Carlo BS (WBS)	2.660	-0.002	0.19
Weighted Monte Carlo H (WH)	2.659	-0.002	0.19
Implied binomial tree (IBT)	3.837	0.063	0.83

3.4.1 American Call Options

Our first analysis deals with the pricing of American index call options.[8] We apply the Black–Scholes formula in combination with the respective estimated volatility for the Black–Scholes model (BS) and the deterministic volatility models (DVF and NTR) to price the American call options. For the Heston model (H), we also use the closed-form pricing formula for European call options with the estimated model parameters. In the implied tree model (IBT), we first determine the cash-flow of each option at the maturity date and work backward through the implied tree afterwards. For the Weighted Monte Carlo approach, we also specify the cash-flow at the maturity date of each option for each simulated path. We generate 50,000 sample paths of (3.1) (WBS), as well as of (3.3) and (3.4), (WH), consisting of 25,000 paths and their antithetics. We then calculate the weighted average over all paths and discount the value to time t_0.

Table 3.6 provides the summary statistics of the out-of-sample performance for the American call options. The pricing errors are generally quite large, at least relative to the in-sample pricing errors reported in Table 3.5. The Black–Scholes model performs worst, as expected, and the implied binomial tree (IBT) approach shows the best out-of-sample pricing performance for the complete sample. These results are also supported by the MOE and MAOPE values. NTR, DVF, H, WBS, and WH show nearly the same figures for all three measures. Interestingly, the Weighted Monte Carlo approaches perform only slightly better than the unweighted prior models, especially for the Heston model. Furthermore, all models underestimate the market prices on average by 15 euro for the whole sample.

We suspect that this underestimation is caused by the differences in the market microstructure between the two exchanges, as well as by the different trading motives of the investors at both markets. The EUWAX caters more toward retail investors with usually speculative motives, while the EUREX is a marketplace for institutional investors with hedging motives.[9] We presume that the 'premium' in the EUWAX quotes is due to differences in transaction costs and hedging costs, as well as in market access restrictions.

---

[8] Since the DAX is a performance index, which is equal to a non-dividend paying underlying, the early exercise opportunity is theoretically worthless. Hence, the price of these American call options should be equal to the European call price with the same strike price and time to maturity.

[9] For a detailed overview of the differences in the market microstructure of the EUREX and EUWAX see Bartram and Fehle (2004).

The RMSE increases with the time to maturity. This holds for all moneyness categories and for all models. This result should be expected since the market price of options is higher for long-term options.

Across all maturities we find somewhat different results for the five moneyness categories. The RMSE increases also from DOTM options to DITM options caused by the higher market prices for DITM options. However, the MAOPE decreases for higher moneyness ratios since the proportion of the time premium at the market price decreases from DOTM options to DITM options. For OTM options, ATM options, and ITM options, the figures are quite similar with a slight tendency to increase from OTM options to ITM options.

The ranking of the models become less clear if we look at individual moneyness maturity groups.[10] All models perform best for short-term DITM options. The main reason for this result is that the theoretical option prices in this class are almost model independent, since they approach the nonparametric lower bound given by the intrinsic value.

For ATM options we find a quite interesting result: The relative performance measured by the MAOPE of the BS model to its alternatives is best for short-term options, whereas all seven models exhibit a similar relative pricing quality for long-term options. This is in line with the common assumption that the BS model is most applicable for short-term options with a strike around the current underlying price (see Hull and White (1987)).

Surprisingly, the NTR performs worst for ATM options independent of the time to maturity, even though it shows the best in-sample performance. This is also the case for ITM long-term options. This result can be explained by the piecewise linear interpolation. For short-term options the volatility structure often must be extrapolated, which increases the pricing error. For long-term options there are in general fewer sampling points, so that the interpolation becomes less precise.

The IBT model performs worst for short-term and mid-term DOTM options measured by the RMSE. This is caused by the discretization of the implied tree and the relatively small number of nonzero states within the tree for these moneyness maturity categories. In contrast to these results, the IBT model performs best for $M > 0.98$, especially for short-term options, since the number of nonzero states increases, which increases the pricing accuracy.

Finally, we take a look at the WMC approaches, in particular, compared to the respective combined model, i.e. BS vs. WBS and H vs.

---

[10] The results for the individual moneyness maturity categories are not reported but are available upon request.

Table 3.6. Out-of-Sample Statistics for the EUWAX American Call Sample

RMSE is the root mean squared euro valuation error. MOE is the mean valuation error. MOE is the mean valuation error outside the observed bid/ask quotes (a positive value indicates that the theoretical value exceeds the ask price on average and a negative value indicates the theoretical value is below the bid price). MAOPE is the median absolute percentage valuation error outside the observed bid/ask quotes (in percent). The sample period is January 2, 2004 to June 30, 2004, with a total of 35,914 observations. BS, NTR, DVF, H, WBS, WH and IBT stand for the Black–Scholes model, the naive-trader-rule, the volatility function approach, the Heston model, the Weighted Monte Carlo approach associated with the Black–Scholes model, the Weighted Monte Carlo approach associated with the Heston model, and the implied binomial tree approach, respectively.

Model	All Options		
	RMSE	MOE	MAOPE
BS	27.39	-15.77	7.13
NTR	25.25	-15.19	6.06
DVF	25.84	-15.39	6.44
H	24.96	-15.63	6.43
WBS	24.74	-15.02	6.07
WH	24.57	-15.08	6.24
IBT	23.54	-13.11	5.30

Model	Time to Maturity								
	Less than 60 days			60 to 180 days			More than 180 days		
	RMSE	MOE	MAOPE	RMSE	MOE	MAOPE	RMSE	MOE	MAOPE
BS	13.09	-6.40	4.70	26.95	-17.24	7.47	44.68	-30.80	10.40
NTR	11.90	-6.59	4.23	23.69	-16.07	6.28	43.14	-30.28	8.90
DVF	11.82	-6.40	4.15	24.48	-16.88	7.02	44.00	-29.60	9.23
H	12.03	-6.78	4.40	24.50	-17.03	6.80	40.77	-29.85	8.82
WBS	11.72	-6.36	4.28	23.56	-15.93	6.36	41.70	-30.16	8.55
WH	11.72	-6.56	4.50	23.38	-15.95	6.46	41.39	-30.07	8.74
IBT	11.26	-5.54	3.68	22.34	-13.18	5.53	39.74	-28.35	8.56

Table 3.6. Out-of-Sample Statistics for the EUWAX American Call Sample – *continued* –

		Moneyness							
Model	$M < 0.94$			$0.94 \leq M < 0.98$			$0.98 \leq M < 1.02$		
	RMSE	MOE	MAOPE	RMSE	MOE	MAOPE	RMSE	MOE	MAOPE
BS	9.15	1.75	16.64	14.01	-6.45	7.69	22.64	-16.50	8.88
NTR	11.12	-4.81	15.91	11.34	-6.30	6.40	24.63	-17.27	9.19
DVF	9.17	-3.12	13.25	14.56	-9.11	8.32	21.47	-15.95	8.32
H	8.72	-3.24	10.61	14.76	-9.31	8.32	21.59	-16.41	8.67
WBS	8.90	-3.72	11.75	11.88	-6.69	6.35	22.42	-16.54	8.55
WH	8.98	-4.22	14.17	11.91	-6.93	6.78	22.41	-16.64	8.68
IBT	9.85	-2.08	12.20	12.13	-5.12	6.06	20.50	-13.73	7.28

		Moneyness				
Model	$1.02 \leq M < 1.06$			$M > 1.06$		
	RMSE	MOE	MAOPE	RMSE	MOE	MAOPE
BS	30.83	-25.51	9.19	36.00	-25.14	3.55
NTR	32.34	-25.44	8.48	31.10	-20.03	2.50
DVF	27.79	-22.29	7.61	34.01	-21.84	2.96
H	27.51	-22.48	7.82	32.46	-22.13	3.05
WBS	30.84	-24.68	8.44	31.35	-20.52	2.69
WH	30.73	-24.53	8.37	31.04	-20.35	2.61
IBT	28.97	-21.65	6.94	29.77	-18.95	2.26

WH. Only for DITM options the WBS and WH model perform consistently better than the respective unweighted model. Roughly speaking, the WBS model slightly improves the BS pricing performance for most moneyness maturity categories. However, for the WH and the Heston model we find opposite results: the Weighted approach performs ten times worse than the original model. This is quite counterintuitive.

A possible explanation for these findings can be that only the underlying prices at the traded maturity dates of each path are incorporated in the optimization algorithm to weight the Monte Carlo paths. Suppose you have options prices for three different maturity dates given with 30, 60, and 90 days to maturity. To price an option with arbitrary maturity of less than 90 days, each generated path in the sample consists of 90 time steps. However, within the calibration algorithm, each path is reduced to three time steps, i.e. it is not ensured that the price process after the calibration is correctly specified for each time step. Another explanation could be that the relative entropy is not an appropriate measure to determine the path probabilities.

3.4.2 European Knock-Out Options

In the second test, we evaluate European knock-out options. For the Black–Scholes model (BS) and the deterministic volatility models, DVF and NTR, we use the closed-form pricing formula derived by Rubinstein and Reiner (1991) in conjunction with the estimated or approximated volatility. The volatility is approximated using the maturity date and the strike price of each knock-out option.[11] In the implied binomial tree model we use standard replication arguments to price the knock-out options.

For the Heston model, as well as for the Weighted Monte Carlo approach, we choose $dt = \frac{1}{365 \times 24}$, i.e. one time step per hour, to approximate the continuous observation of the barrier. Then, we generate 50,000 sample paths of (3.1), as well as of (3.3) and (3.4), consisting of 25,000 paths and their antithetics. For the Heston model we take the simple mean of all paths as a proxy for the model price, whereas for the Weighted Monte Carlo approach we calculate the weighted average over all paths.

The summary statistics of the out-of-sample pricing performance are shown in Tables 3.7 and 3.8 for the down-and-out calls and the up-and-out puts, respectively. In the following consideration, we will not dwell

---

[11] Note that the maturity date is only a proxy for the time to maturity or time to knock-out, since we do not know the time of a possible knock-out.

on the results of the long-term options, since there is an insufficient number of observations.

Looking at MAOPE in Table 3.7 first, we find that the non-parametric pricing models, WH, WBS, and IBT, perform best over all observations and that the Black–Scholes (BS) model and the deterministic models, DVF and NTR, perform worst, with the Heston (H) in between the two extremes. It is quite surprising that the IBT model has the best performance for knock-out options, since Boyle and Lau (1994) have illustrated that the binomial option pricing algorithm can lead to significantly biased estimates in the prices for this option type. We presume that due to the special semi-recombining structure of the implied binomial tree this bias becomes less pronounced.

At a first glance, it seems as if the pricing performances were better here than for the American call options. However, this is mainly a result of the missing DOTM and OTM options with the highest MAOPE values, since these are options that are knocked-out. The result also holds over all moneyness categories for the respective maturity class.

If we look at individual moneyness maturity categories[12] it becomes apparent, that the Black–Scholes (BS) model, NTR, and DVF perform virtually equally. This could be caused by the application of the closed-form pricing formula derived by Rubinstein and Reiner (1991). Similar results are reported by Andersen and Brotherton-Ratcliffe (1998). They show that knock-out prices can be bounded, nonmonotonic functions of implied volatilities. Therefore, they recommend applying the concept of implied volatility for knock-out options with care. Our results support this recommendation. This holds for the down-and-out call options, as well as for the up-and-out put options.

For the ATM options, the RMSE and the MAOPE are the highest across all options. This is due to the fact that the current index level is close to the strike price (barrier). Therefore, the risk for a knock-out is quite high, so that the intrinsic value goes down. The option price is mainly dominated by the time premium; thus, the scope for mispricing is quite high. For this moneyness category, the IBT and Heston (H) model performs best.

For the Weighted Monte Carlo approaches we find similar results as for the American call options. The WH model performs worse than the original Heston model for ATM and ITM calls and only slightly better for DITM down-and-out calls. The WBS model halves the MAOPE for DITM down-and-out calls as compared with the original BS model. It

---

[12] The results for the individual moneyness maturity categories are not reported but are available upon request.

Table 3.7. Out-of-Sample Statistics for the EUWAX Down-and-Out Call Sample

RMSE is the root mean squared euro valuation error. MOE is the mean valuation error. MOE is the mean valuation error outside the observed bid/ask quotes (a positive value indicates the theoretical value exceeds the ask price on average; a negative value indicates the theoretical value is below the bid price). MAOPE is the median absolute percentage valuation error outside the observed bid/ask quotes (in percent). The sample period is January 2, 2004 to June 30, 2004, with a total of 9,026 observations. BS, NTR, DVF, H, WBS, WH, and IBT stand for the Black–Scholes model, the naive-trader-rule, the volatility function approach, the Heston model, the Weighted Monte Carlo approach associated with the Black–Scholes model, the Weighted Monte Carlo approach associated with the Heston model, and the implied binomial tree approach, respectively.

| Model | All Options | | | Time to Maturity | | | | | | | | |
| | | | | Less than 60 days | | | 60 to 180 days | | | More than 180 days | | |
	RMSE	MOE	MAOPE	RMSE	MOE	MAOPE	RMSE	MOE	MAOPE	RMSE	MOE	MAOPE
BS	15.44	-8.50	2.56	17.82	-9.05	2.77	9.00	-7.19	2.20	17.05	-13.40	2.08
NTR	15.54	-8.68	2.65	17.87	-9.14	2.82	9.25	-7.48	2.33	17.94	-14.55	2.35
DVF	15.52	-8.64	2.63	17.87	-9.13	2.81	9.22	-7.44	2.29	17.46	-13.89	2.25
H	14.38	-6.79	2.05	16.95	-7.64	2.36	6.98	-4.93	1.57	14.73	-10.88	2.13
WBS	14.36	-4.42	1.48	16.72	-5.91	1.57	8.06	-1.76	1.31	12.90	1.08	1.44
WH	14.23	-6.08	1.80	16.75	-7.05	2.04	6.92	-4.04	1.26	15.78	-9.26	1.78
IBT	13.32	-2.30	1.28	15.74	-3.94	1.46	6.62	0.87	0.95	9.81	-1.72	0.81

| Model | Moneyness | | | | | | | | |
| | $0.98 \leq M < 1.02$ | | | $1.02 \leq M < 1.06$ | | | $M \geq 1.06$ | | |
	RMSE	MOE	MAOPE	RMSE	MOE	MAOPE	RMSE	MOE	MAOPE
BS	40.45	-17.23	16.68	16.50	-9.45	5.56	9.02	-7.22	1.79
NTR	40.42	-17.17	16.66	16.54	-9.53	5.60	9.26	-7.46	1.86
DVF	40.46	-17.26	16.70	16.58	-9.59	5.65	9.18	-7.38	1.84
H	38.82	-12.64	9.67	14.72	-6.17	3.75	8.34	-6.44	1.54
WBS	40.13	-16.15	16.40	15.68	-6.68	4.01	6.81	-2.30	0.94
WH	38.93	-13.44	11.05	15.27	-7.07	4.28	7.46	-4.93	1.11
IBT	37.51	-1.41	9.34	13.98	1.55	2.36	6.64	-3.96	0.96

Table 3.8. Out-of-Sample Statistics for the EUWAX Up-and-Out Put Sample

RMSE is the root mean squared euro valuation error. MOE is the mean valuation error outside the observed bid/ask quotes (a positive value indicates the theoretical value exceeds the ask price on average; a negative value indicates the theoretical value is below the bid price). MAOPE is the median absolute percentage valuation error outside the observed bid/ask quotes (in percent). The sample period is January 2, 2004 to June 30, 2004, with a total of 6,848 observations. BS, NTR, DVF, H, WBS, WH, and IBT stand for the Black–Scholes model, the naive-trader-rule, the volatility function approach, the Heston model, the Weighted Monte Carlo approach associated with the Black–Scholes model, the Weighted Monte Carlo approach associated with the Heston model, and the implied binomial tree approach, respectively.

| Model | All Options | | | Time to Maturity | | | | | | | | |
| | | | | Less than 60 days | | | 60 to 180 days | | | More than 180 days | | |
	RMSE	MOE	MAOPE	RMSE	MOE	MAOPE	RMSE	MOE	MAOPE	RMSE	MOE	MAOPE
BS	22.67	-13.28	4.78	24.61	-11.86	4.13	16.37	-13.93	6.16	37.21	-34.83	7.24
NTR	22.83	-13.48	4.87	24.67	-11.99	4.22	16.60	-14.18	6.22	38.53	-35.92	7.58
DVF	22.81	-13.48	4.87	24.66	-11.97	4.20	16.69	-14.27	6.27	37.78	-35.31	7.31
H	21.50	-11.47	3.96	23.72	-10.39	3.50	14.39	-11.53	4.90	35.62	-32.73	6.80
WBS	25.28	-16.29	6.11	25.95	-14.50	5.24	20.37	-17.06	7.06	49.41	-43.96	8.64
WH	22.37	-12.63	4.49	24.37	-11.74	4.10	15.04	-12.14	5.07	40.63	-36.22	6.83
IBT	19.20	-6.59	2.33	21.90	-6.61	2.28	10.34	-4.73	2.23	32.13	-27.00	4.77

| Model | Moneyness | | | | | | | | |
| | $M < 0.94$ | | | $0.94 \leq M < 0.98$ | | | $0.98 \leq M < 1.02$ | | |
	RMSE	MOE	MAOPE	RMSE	MOE	MAOPE	RMSE	MOE	MAOPE
BS	16.13	-12.95	2.92	17.01	-11.57	6.82	49.10	-20.80	17.32
NTR	16.55	-13.27	2.98	17.07	-11.68	6.92	49.10	-20.80	17.25
DVF	16.42	-13.21	2.98	17.13	-11.73	6.92	49.11	-20.85	17.39
H	15.35	-12.12	2.70	15.21	-9.07	5.24	47.59	-17.22	11.48
WBS	22.31	-18.40	4.23	18.56	-13.13	7.95	47.97	-18.43	14.41
WH	17.18	-13.64	3.09	15.66	-9.84	5.85	47.90	-18.22	13.39
IBT	13.21	-9.36	2.04	12.18	-2.78	2.42	44.64	-8.22	7.55

generally performs better than BS for short-term options but for mid-term options we find mixed results.

The MOE is mostly negative for down-and-out calls and up-and-out puts for all models, which indicates an underestimation of the market quotes at EUWAX. Again, this is caused by the differences in market microstructure of the EUWAX exchange as a market for retail investors and is not mainly due to the models.

We find predominantly the same qualitative statements for the out-of-sample performance of up-and-out put options (keep in mind that put options are DITM for ($M < 0.94$) and ITM for ($0.94 \leq M < 0.98$)). The main exception is the performance of the WBS model for up-and-out put options. Surprisingly, this model performs the worst for this option type. This supports our hypothesis in the previous section about the specification of the price process after weighting the sample paths.

Roughly speaking the RMSE, MOE, and MAOPE are twice as high for the up-and-out put options as compared with the down-and-out call options. Only for the short-term ITM and short-term DITM puts do the figures remain on the same level.

3.5 Conclusion

Under no-arbitrage conditions the price for the same contingent claim should be equal, independent of where the claim is traded. Starting with this assumption, we compared the Black–Scholes model, the Heston model; the Weighted Monte Carlo approach combined with both models as proposed by Avellaneda et al. (2001); the implied binomial tree model suggested by Herwig (2005); and two deterministic volatility models, namely the specification of a volatility function and a naive-trader-rule with respect to their out-of-sample performance. For this objective we calibrate these models to market prices of European call options traded at the EUREX. Afterwards we price American call options, down-and-out call options, and up-and-out put options with these calibrated models and compare the theoretical value to market quotes for these options from the EUWAX. This study presents the first empirical test of the Weighted Monte Carlo approach suggested by Avellaneda et al. (2001) and the implied tree model developed in Herwig (2005).

For the overall performance, it turns out that the implied tree model performs best for all observed option categories. As expected, the Black–Scholes model shows the worst pricing performance for American calls and down-and-out call options. For these two option types the Weighted

Monte Carlo approaches (WBS and WH) perform quite well. Surprisingly for us, the Weighted Monte Carlo approach combined with the Black–Scholes model (WBS) has the worst pricing performance for up-and-out put options. We also find another phenomenon, namely that the 'weighted models', WBS and WH, perform worse than the original models, BS and H, in some moneyness maturity categories. We suspect that this is caused by the weighting of the sample paths, which then influences the price process of the underlying. The performance of the deterministic volatility models and the Heston model is mixed.

For the pricing of the knock-out options we apply the closed-form pricing formula derived by Rubinstein and Reiner (1991). In accordance with Andersen and Brotherton-Ratcliffe (1998) we find that this pricing-formula should only be used with care when combined with the concept of implied volatilities.

Another interesting observation is that generally all models for all option types underestimate the quoted prices for the options at the EUWAX. We suspect that these prices contain a premium, since at the EUWAX mainly retail investors act as traders, whereas the EUREX is dominated by institutional investors.

Future research should examine whether EUWAX traded options really exhibit a premium, and, if so, how it could be specified and explained. Since there are different investment banks that issue options at the EUWAX, it would be interesting to see if there were differences between the issuers. Another research question is the specification of the Weighted Monte Carlo approach. On the one hand, it should be examined if additional benchmark instruments, e.g. forwards or futures, improve the pricing performance. On the other hand, different criteria to determine the path probabilities should be compared.

References

Aït-Sahalia, Y. and A.W. Lo (1998): Nonparametric Estimation of State-Price Densities Implicit in Financial Asset Prices, *The Journal of Finance*,
53(2):499–547.

Andersen, L.B.G. and R. Brotherton-Ratcliffe (1998): The Equity Option Volatility Smile: An Implicit Finite-Difference Approach, *The Journal of Computational Finance*, 1(2):5–37.

Avellaneda, M., R. Buff, C. Friedman, N. Grandchamp, L. Kruk, and J. Newman (2001): Weighted Monte Carlo: A New Technique for Calibrating Asset-Pricing Models, *International Journal of Theoretical and Applied Finance*, 4(1):91–119.

Bakshi, G., C. Cao, and Z. Chen (1997): Empirical Performance of Alternative Option Pricing Models, *The Journal of Finance*, 52(5):2003–2049.

Barle, S. and N. Cakici (1998): How to Grow a Smiling Tree, *Journal of Financial Engineering*, 7(2):127–146.

Bartram, S.M. and F.R. Fehle (2004): *Alternative Market Structures for Derivatives*, EFA 2003 Annual Conference Paper No. 297.

Belledin, M. and C. Schlag (1999): An Empirical Comparison of Alternative Stochastic Volatility Models, *Working Paper Series Finance & Accounting, Fachbereich Wirtschaftswissenschaften*, Goethe–University, Frankfurt am Main

Black, F. and M. Scholes (1973): The Valuation of Options and Corporate Liabilities, *Journal of Political Economy*, 81(3):637–654.

Boyle, P. and S.H. Lau (1994): Bumping Up Against the Barrier with the Binomial Method, *The Journal of Derivatives*, 1(4):6–14.

Brown, G. and K.B. Toft (1999): Constructing Binomial Trees from Multiple Implied Probability Distributions, *The Journal of Derivatives*, 7(2):83–100.

Derman, E. and I. Kani (1994): Riding on a Smile, *RISK*, 7(2):32–39.

Derman, E., I. Kani, and N. Chriss (1996): Implied Trinomial Trees of the Volatility Smile, *The Journal of Derivatives*, 3(4):7–22.

Dumas, B., J. Fleming, and R.E. Whaley (1998): Implied Volatility Functions: Empirical Tests, *The Journal of Finance*, 53(6):2059–2106.

Dupire, B. (1994): Pricing with a Smile, *RISK*, 7(1):18–20.

Herwig, T. (2005): Construction of Arbitrage-Free Implied Trees: A New Approach, Working Paper, Goethe-University, Frankfurt am Main, *http://ssrn.com/abstract=643584*.

Heston, S. (1993): A Closed-Form Solution for Options with Stochastic Volatility with Applications to Bond and Currency Options, *The Review of Financial Studies*, 6(2):327–343.

Hull, J.C. and A. White (1987): The Pricing of Options on Assets with Stochastic Volatilities, *The Journal of Finance*, 42(2):281–300.

Jackwerth, J.C. (1997): Generalized Binomial Trees, *The Journal of Derivatives*, 5(2):7–17.

Jackwerth, J.C. and M. Rubinstein (1996): Recovering Probability Distributions from Option Prices, *The Journal of Finance*, 51(5):1611–1631.

Jackwerth, J.C. and M. Rubinstein (2001):   Recovering Stochastic
Processes from Option Prices,   *Working Paper*, London Business
School.

Rubinstein, M. (1994): Implied Binomial Trees, *The Journal of Finance*,
49(3):771–818.

Rubinstein, M. and E. Reiner (1991):   Breaking Down the Barriers,
RISK,
4(8):28–35.

Tompkins, R.G. (2001a): Implied Volatility Surfaces: Uncovering Reg-
ularities for Options on Financial Futures, *The European Journal of
Finance*, 7(3):198–230.

Tompkins, R.G. (2001b): Stock Index Futures Markets: Stochastic Volatility Models and Smiles, *The Journal of Futures Markets*, 21(1),43–
78.

4

Market-Conform Valuation of American-Style Options via Monte Carlo Simulation

4.1 Introduction

Two specific problems in option pricing theory are the calibration of models to given market prices and the valuation of options with American-style exercise features. The most prominent approaches to handle the combination of both problems are implied trees (e.g. Rubinstein (1994), Dupire (1994), Derman and Kani (1994), Jackwerth (1997), or Brown and Toft (1999)). These models are consistent with the market volatility smile and can be used to price standard American options. However, they are not well-suited for the pricing of more complex derivatives, such as Asian options, lookback options, basket options, or options on multiple assets. The pricing becomes even more complex when the options include American-style exercise features. One alternative solution for the valuation of complex derivatives in high-dimensional models are simulation techniques. Different approaches, usually based on Monte Carlo simulation, exist for the pricing of American options and the calibration to given market prices in high dimensional models. However, these approaches deal separately with the two problems mentioned above. Therefore, we propose two new flexible approaches to price American-style options in accordance with given market prices via Monte Carlo simulation.

Usually lattice methods or finite differences are used to price American-style options. These techniques work backwards from the maturity date of the option. However, they are inefficient for higher dimensional problems (i.e. problems with more than three state variables), and they are very difficult to apply to path-dependent options. Simulation techniques are often used here, since they are simple and flexible. However, for a long time, simulation techniques seemed inapplicable to American-style options. This is due to their forward-construction principle and their path-by-path generation. Since American-style contingent claims are traded in all important derivative markets, many suggestions have been made during the last years to price American-style options by simulation and, in particular, to determine the optimal early-exercise strategy. Examples can be found in Tilley (1993), Barraquand and Martineau (1995), Carriere (1996), Broadie and Glasserman (1997), Andersen (1999), Longstaff and Schwartz (2001), or Ibánez and Zapatero (2004).

Beyond this, there is another common problem in asset pricing theory. In efficient markets, prices of actively traded securities should be equal to their expected discounted cash-flows under the risk-neutral measure for given model parameters. Thus, it is assumed that the parameters of the pricing model are known. However, in empirical applications this is typically not the case, and the parameters have to be estimated. The so-called 'calibration of asset price models' uses the information contained in market prices in such a way that the pricing model reproduces the prices of all benchmark instruments correctly. Unfortunately, in most of the cases, such a correct model specification is not possible due to the individual model assumptions. For example, the classical Black–Scholes model assumes a constant volatility of the underlying, which is the only unknown parameter in the model. Many empirical studies (e.g. Rubinstein (1994), Jackwerth and Rubinstein (1996), Dumas et al. (1998), and Aït-Sahalia and Lo (1998)), however, document the existence of a systematic relationship between the implied volatility and the strike price or moneyness of an option, known as the volatility smile. Therefore, it is not possible to estimate a constant volatility parameter that would simultaneously price all traded options correctly.

For high-dimensional models, the parameter estimation is quite challenging, since frequently no closed-form pricing formula is available. Avellaneda et al. (2001) present a new approach for calibrating Monte Carlo simulations to the market prices of benchmark securities. Starting from a given model of market dynamics, they correct misspecifications in the simulation by assigning new 'probability weights' to each simulated path.

In order to solve both above mentioned problems simultaneously, we propose two new intuitive and efficient valuation methods for American-style options via Monte Carlo simulation in accordance with the given market prices of actively traded benchmark instruments. Our first approach combines the Weighted Monte Carlo technique by Avellaneda et al. (2001) with the approach by Andersen (1999), while the second method merges the Weighted Monte Carlo method with the Least-Squares Monte Carlo approach by Longstaff and Schwartz (2001). In addition, we show the effect of weighting Monte Carlo simulations. Afterwards, we apply the original approaches by Andersen (1999) and Longstaff and Schwartz (2001), as well as our two extensions, to real market data from the EUREX and EUWAX exchanges to investigate the performance of the different methods.

The chapter is structured as follows. Section 4.2 describes the Monte Carlo method to price derivatives. We present our two extensions to price American-style options by simulation in accordance with given market prices in Sect. 4.3. Section 4.4 shows the effect of weighting Monte Carlo simulations. Section 4.5 contains a comparison of the pricing of American put options with and without weighting the sample paths for different pricing models. The final section concludes with a summary of our results and an outlook.

4.2 Monte Carlo Methods

4.2.1 The Classical Monte Carlo Technique

Before we go into a detailed description of the Weighted Monte Carlo method proposed by Avellaneda et al. (2001), we will briefly review the classical Monte Carlo method. In general, the classical Monte Carlo approach to estimate the value of an option by simulation consists of the following steps:

1. Simulation of the sample paths of the underlying state variables (e.g. underlying asset prices, volatility, or interest rates) under the risk-neutral measure[1] over the relevant time horizon by using a pseudo-random numbers generator.
2. Calculation of the discounted payoff of the security on each sample path dependent on the structure of the security.
3. Calculation of the average discounted payoff over all sample paths.

The Monte Carlo approach becomes more attractive as compared with finite difference methods or binomial techniques as the dimension of

---

[1] Under the assumption of no arbitrage, the price of a common derivative security can be expressed as the expected value of its discounted payoffs under the risk-neutral measure. For a detailed description see, for example, Musiela and Rutkowski (1997) or Duffie (2001).

the problem increases. This is because it can easily deal with multiple random factors, such as options on multiple assets, stochastic volatility, or path dependent options. The strong law of large numbers ensures that the estimate converges almost surely to the true value of the option and the central limit theorem ensures that the standard error of the estimate tends to zero at rate $\frac{1}{\sqrt{M}}$, where M represents the number of simulated paths.[2] For less than three dimensions, the Monte Carlo simulation is inefficient compared to finite difference methods, since it is computationally more expensive to estimate with reasonable accuracy. To speed up convergence several methods of variance reduction exist, e.g. antithetic variables, control variates, moment matching methods, or stratified sampling. See Boyle et al. (1997) or Wilmott (1999) for a detailed description of these methods.

Further benefits of using Monte Carlo methods are that the mathematics are very basic, that correlations can be modelled easily, that the error bound is independent of the dimension, and that the method is very flexible. In addition, the increased availability of more powerful computers and software packages in recent years makes it easier to implement the method and enhances its attractiveness. The main disadvantage is that the method is slow as compared with the finite difference method to solve partial differential equations for problems up to three or four dimensions and provides only a probabilistic error bound.

4.2.2 Weighted Monte Carlo Technique

The key idea of the Weighted Monte Carlo approach by Avellaneda et al. (2001) is to precisely replicate spot prices of liquid and actively traded options via Monte Carlo simulation. Simulation in the classical Monte Carlo framework requires an estimation of the unknown parameters of the stochastic processes of the state variables. Therefore, one has to calibrate the model to the market prices of the benchmark instruments by inverting the corresponding pricing formula. One problem with regard to estimating the parameters is that closed-form solutions of pricing formulas are not available for each benchmark instrument. Another way of estimating the unknown parameter involves real statistics, such as estimates of rates of return, historical volatilities, or correlations. Nevertheless, the correct pricing of all benchmark instruments simultaneously generally fails.

---

[2] See Niederreiter (1992) for an excellent discussion of error bounds in Monte Carlo methods.

The approach by Avellaneda et al. (2001) uses another, non-parametric solution to price the securities via simulation. Here the objective is to determine the risk-neutral probabilities of the future states of the market directly instead of searching for the parameters of the stochastic processes to generate the sample paths for the state variables. In contrast to the classical approach, they use 'weighted' paths instead of the average over all sample paths to determine the option value. The probabilities p_i for each path are determined such that:

- The expected values of the discounted cash-flows of the benchmark instruments minimize the weighted squared pricing error, i.e.

$$\min_p \chi_w^2 = \frac{1}{2} \sum_{j=1}^{O} \frac{1}{w_j} (\mathbb{E}^p[g_j] - C_j)^2. \tag{4.1}$$

C_j is the market price of benchmark instruments, O is the number of benchmark instruments, χ_w^2 is the sum of weighted least-squares residuals, and w_j are positive weights. The limit $w_j \rightarrow 0$ corresponds to an exact matching of market prices. The relaxation of exact matching is due to bid-ask-spreads and liquidity considerations and is recommended to avoid numerical problems.
- They are as close as possible to the prior probabilities q_i, i.e. they minimize the relative entropy distance $D(p|q)$, where

$$D(p|q) = \sum_{i=1}^{M} p_i \log \left(\frac{p_i}{q_i} \right), \tag{4.2}$$

and where M is the number of paths. For Monte Carlo simulations we have uniform probabilities $u = q_i \equiv \frac{1}{M}$, and $D(p|q)$ can be written as

$$D(p|u) = \log(M) + \sum_{i=1}^{M} p_i \log(p_i). \tag{4.3}$$

Avellaneda et al. (2001) propose the *Kullback–Leibler relative entropy* to find the 'calibrated' probabilities.[3] The relative entropy measures the deviation of the calibrated model from the prior. Therefore, $D(p|u)$ has to be minimized to receive the minimal distance from the prior. The algorithm to calibrate Monte Carlo simulations under market price constraints consists of the following steps:

---

[3] Entropy theory was introduced in finance theory for single-period models by Buchen and Kelly (1996) and Gulko (1999, 2002).

i. Construct a set of sample paths under the risk-neutral measure using the following stochastic difference equation and a pseudo-random number generator

$$\mathbf{X}_{t_{k+1}} = \mathbf{X}_{t_k} + \sigma(\mathbf{X}_{t_k}, t_k)\boldsymbol{\xi}_{t_{k+1}}\sqrt{\Delta t} + \boldsymbol{\mu}(\mathbf{X}_{t_k}, t_k)\Delta t, \quad k = 0, ..., N-1,$$

(4.4)

where T_{max} is the time horizon, N represents the number of discrete time steps, $\mathbf{X}_{t_k} \in \mathbb{R}^d$ is a vector of state variables, and $\boldsymbol{\xi}_{t_k} \in \mathbb{R}^{d'}$ is a vector of independent Gaussian shocks, which are $IIN \sim (0, 1)$. The $d \times d'$ matrix $\sigma(\mathbf{X}_{t_k}, t_k)$ represents the variance-covariance structure, and the drift is represented by the d-vector $\boldsymbol{\mu}(\mathbf{X}_{t_k}, t_k)$. The drift-parameter and the variance-covariance parameter should be estimated as described above.

ii. Calculate the discounted cash-flow matrix $\mathbf{g} \in \mathbb{R}^{M \times O}$ for the benchmark instruments. Benchmark instruments can be forwards, futures, standard European options, path-dependent options, bonds, swaps, and caps. American-style derivatives cannot be used to calibrate the model, since it is not possible to determine the discounted cash-flow for this type of options.

iii. Using (4.1) and (4.3) the following optimization problem needs to be solved:

$$\min_{p} \left\{ D(p|u) + \chi_w^2 \right\}$$

(4.5)

$$s.t. \sum_{i=1}^{M} p_i = 1,$$

(4.6)

$$p_i \geq 0 \quad \forall i.$$

(4.7)

Rewriting (4.5) leads to:[4]

$$- \inf_{\lambda} \left\{ \log(Z(\lambda)) - \sum_{j=1}^{O} \lambda_j C_j - \log(M) + \frac{1}{2}\sum_{j=1}^{O} w_j \lambda_j^2 \right\},$$

(4.8)

which can be solved with a gradient-based optimization routine. The normalization factor $Z(\lambda)$ ensures that constraint (4.6) is fulfilled. The solution of this optimization is given in (4.9) and (4.10).

iv. Compute the risk-neutral probabilities p_i, $i = 1, ..., M$ for each path using the optimal values of $\lambda_1, ..., \lambda_O$:

$$\lambda_j^* = -\frac{1}{w_j}(\mathbb{E}^{p^*}[g_j] - C_j)$$

(4.9)

---

[4] See Avellaneda et al. (2001) for details of the transformations.

$$p_i^* = \frac{1}{Z(\lambda^*)} \exp\left(\sum_{j=1}^{O} \lambda_j^* g_{ij}\right). \tag{4.10}$$

The Weighted Monte Carlo (WMC) approach uses market data in two steps. The first step incorporates market-implied and/or historical data to generate sample paths. This represents a first guess for the prior 'risk-neutral' probability measure. The second step re-calibrates the prior information to the observed market prices. As we will show in Sect. 4.4, it is highly recommended to use the best available guess for the unknown parameters, since there is only limited possibility for weighting the sample paths.

The algorithm can be used for stochastic volatility models, stochastic interest rate models, or jump-diffusion models. Furthermore, Avellaneda et al. (2001) show that the method implicitly uses control variates. This leads to a significant variance reduction, especially for options with the same maturity as the benchmark instruments.

4.3 Valuing American Options by Simulation

4.3.1 Weighted Threshold Approach

Andersen (1999) proposes a simple method to price Bermudan swaptions in the LIBOR market model. This model class allows for the incorporation of multiple stochastic factors and volatility smiles and has enough degrees of freedom to fit the market prices of interest rate derivatives as well. Therefore, the model has a high number of state variables, which requires Monte Carlo methods to price derivatives. The early exercise strategy for American-style options depends on all these state variables, which leads to a high-dimensional optimization problem. To reduce the complexity of the problem, Andersen (1999) proposes a one-dimensional optimization by maximizing the value of the option at each time step, searching the early exercise boundary parameterized in intrinsic value and the value of still-alive swaptions. This is done by introducing a Boolean function with a single, time-dependent parameter, which is determined in the optimization routine for each time step. As a result, the method is simple to implement, fast and robust, and produces a lower bound for Bermudan swaption prices.

Douady (2001) shows that the objective function in the optimization routine is not smooth. Therefore, not every algorithm to maximize a function of one variable guarantees convergence to the true maximum.

Hence, Douady (2001) extends Andersen's (1999) approach by introducing 'exercise probabilities', which depends on the fuzziness parameter α and leads to a 'fuzziness' of the threshold. Thereby, the new objective function is smooth enough for an appropriate 'fuzziness' parameter α to identify the maximum of the function uniquely.[5]

The method of Andersen (1999) can also be easily applied to other model classes, especially to equity models. In this case, only the market variables change but both the algorithm and the general idea remain unchanged. As a result, another problem emerges that is not present in the LIBOR market model, namely good model calibration to market prices of plain-vanilla options. Therefore, we propose to merge the methods suggested by Andersen (1999) and by Avellaneda et al. (2001) to mitigate these problems. We refer the resulting algorithm as the 'Weighted Threshold Approach' (WTA). This algorithm to price American-style options contains of the following steps:

- Calibrate the simulated paths to the market prices of benchmark instruments as described in Sect. 4.2.2 by steps i. to iv. to get the path probabilities p_i^*. Use the same setting and parameters also in the following steps.
- Starting at t_{N-1}, maximize the value of the option by searching for the optimal threshold θ_k for each discrete time point $k = 1, ..., N - 1$ using backward induction and solve the following optimization problem:

$$\max_{\theta_k} \sum_{i=1}^{M} p_i^* \left\{ IV_{i,k} f(\theta_k) + H_{i,k}(1 - f(\theta_k)) \right\}. \tag{4.11}$$

Here $IV_{i,k}$ denotes the intrinsic value of the option at path i for immediate exercise at time t_k, and $H_{i,k}$ is the value if the option is not exercised at that date. Using the Andersen (1999) approach

$$H_{i,k} = \exp\left(-r(t_i^* - t_k)\right) IV_{i,t_i^*} \tag{4.12}$$

where t_i^* represents the current optimal exercise date at path i and

$$f(\theta_k) = \mathbf{1}_{\{IV_{i,k} \geq \theta_k\}} \tag{4.13}$$

is the indicator function which controls the optimal exercise at t_k. In order to keep the problem simple, we assume, without loss of

---

[5] A non-fuzzy threshold corresponds to $\alpha \to +\infty$. The choice of α should be tested numerically depending on the functional form of the exercise probabilities. For details see Douady (2001).

generality, a constant risk-free interest rate r. In the extended version of Douady (2001) the holding value $H_{i,k}$ and the exercise probability $f(\theta_k)$ are defined as

$$H_{i,k} = \exp\left(-r(t_{k+1} - t_k)\right)\left[IV_{i,k+1}f(\theta_{k+1}) + H_{i,k+1}(1 - f(\theta_{k+1}))\right] \tag{4.14}$$

and

$$f(\theta_k) = \frac{1}{1 + \exp\left(-\alpha(IV_{i,k} - \theta_k)\right)}, \tag{4.15}$$

where α measures the fuzziness of the threshold. Note that the smoothing in the extended version is due to the replacement of the indicator function by a continuous function in θ_k.

- Use the completely specified exercise strategy $\Theta = (\theta_1, ..., \theta_{N-1})$ to price the American-style option by an independent Monte Carlo simulation with $\hat{M} \gg M$ paths. As a consequence of the higher number of sample paths, the path probabilities p_i^* changes. Hence, the calibration algorithm from Sect. 4.2.2 must also be repeated.

The proposed method keeps the advantages of the original method from Andersen (1999), since for $p_i \equiv \frac{1}{M}$, $i = 1, ..., M$ the method exactly corresponds to Andersen's approach. The major drawback of the extended algorithm is that the Weighted Monte Carlo technique must be done twice, which can be computationally time-consuming.

4.3.2 Weighted Least-Squares Approach

Longstaff and Schwartz (2001) suggest to estimate the continuation value $F(i, t_k)$ as the conditional expectation of the payoff from keeping the option alive, using the cross-sectional information in the simulation. Again, i represents a sample path and t_k are the discrete time points at which the American option can be exercised. Longstaff and Schwartz (2001) assume a complete probability space $(\Omega, \mathcal{F}, P)$, a finite time horizon $[0, T]$, and the existence of an equivalent martingale measure Q for the economy. They restrict their attention to derivatives with payoffs in $L^2(\Omega, \mathcal{F}, Q)$. Hence, the conditional expectation is also in $L^2(\Omega, \mathcal{F}, Q)$. Since L^2 is a Hilbert space any function belonging to this space can be represented as a countable linear combination of $\mathcal{F}_{t_k}$-measurable basis functions for this vector space. Longstaff and Schwartz (2001) choose Laguerre polynomials to represent the basis functions[6]

---

[6] Other types of basis functions include the Hermite, Legrende, Chebyshev, Gegenbauer, or Jacobi polynomials. For Hilbert space theory see Heuser (1992) or Achieser and Glasmann (1981).

$$L_n(X) = \exp\left(-X/2\right) \frac{e^X}{n!} \frac{d^n}{dX^n}(X^n e^{-X}), \quad n = 0, ..., \tag{4.16}$$

where L_n defines the Laguerre polynomial of order n at point X. With this specification the continuation value on path i can be written as

$$F(i, t_k) = \sum_{j=0}^{\infty} \beta_j L_j(X). \tag{4.17}$$

In order to use this result in practice one needs to approximate $F(i, t_k)$ using a finite linear combination with the first K $< \infty$ basis functions. We denote this approximation as $F_K(i, t_k)$. A natural approximation concept for $F_K(i, t_k)$ is that of least-squares regression, where the discounted cash-flow $CF_{i,t_{k+1}}$ one period ahead is projected on a set of basis functions for the paths where the option is 'in-the-money' at time t_k.[7]

The Longstaff and Schwartz (2001) approach is a simple technique to approximate the value of American-style options. It is easy to implement, since only the least-squares regression and simulation techniques are required. The disadvantage of the approach is that there is no unique procedure how to specify the structure and the form of the basis functions. Nonetheless, it is a very powerful pricing method, which values American-style options accurately by simulation for given parameter specification of the pricing model, e.g. the Black–Scholes model.

However, there is no suggestion how this method can be related to market prices and no results about its performance under real market conditions are available. Therefore, we link the Longstaff and Schwartz (2001) approach to the Weighted Monte Carlo method to price American-style derivatives in accordance with market prices of benchmark instruments. The critical point is that we must take into account that the paths have different weights. Hence, we use weighted least-squares[8] to estimate the regression coefficients β_j in (4.17). Consequently, we call the new algorithm 'Weighted Least-Squares Approach' (WLSA). We are able to price American-style derivatives under the 'calibrated' risk-neutral measure.

The resulting algorithm works as follows:

- Calibrate the simulated paths to the market prices of benchmark instruments as described in Sect. 4.2.2 by steps i. to iv. to get the

---

[7] Using only 'in-the-money' paths allows a better estimate of the conditional expectation function in the relevant exercise region and improves the efficiency significantly.

[8] For an overview see Greene (1999) or Johnston and DiNardo (1997).

path probabilities p_i^*. Use the same setting and parameters also in the following steps.

- Work recursively and use at any time t_k, $0 < k < N$ weighted least-squares regression to estimate the conditional expectation of the payoff $F_K(i, t_k)$, given that the option is kept alive. Construct a subset of $\widehat{M}$ paths consisting of paths where the option is in-the-money. The regression coefficients $\tilde{\boldsymbol{\beta}}$ are given by

$$\tilde{\boldsymbol{\beta}} = (\mathbf{A}'\mathbf{W}\mathbf{A})^{-1}\mathbf{A}'\mathbf{W}\mathbf{y} \tag{4.18}$$

where the $\widehat{M}$-vector $\mathbf{y}$ represents the current holding values given as

$$y_i = \exp\left(-r(t_i^* - t_k)\right) CF_{i,t_i^*}. \tag{4.19}$$

Here t_i^* denotes the current optimal exercise time for the option on the $i-th$ path. The $\widehat{M} \times (K+2)$-matrix $\mathbf{A}$ contains the basis functions with $a_{i,n} = L_n\left(X_{i,t_k}^{(m)}\right)$ for $n = 0, ..., K$, where $X_{i,t_k}^{(m)}$ denotes the $m-th$ state variable at time t_k on the $i-th$ path and $a_{i,K+1} = 1$.[9] The diagonal-elements $w_{i,i}$ of the $\widehat{M} \times \widehat{M}$-matrix $\mathbf{W}$ are occupied with the 'in-the-money' path probabilities p_i^*. All other elements are zero.[10]

Then, the fitted values $\hat{\mathbf{y}} = \mathbf{A}\tilde{\boldsymbol{\beta}}$ are used as a proxy for the continuation value. Hence, it is easy to decide at time t_k, whether it is optimal to exercise the option at path i. If the value of immediate exercise is greater than the continuation value, set CF_{i,t_i^*} equal to the immediate exercise value and $t_i^* = t_k$, otherwise leave CF_{i,t_i^*} and t_i^* unchanged.

- Once t_0 is reached and the optimal exercise strategy is determined, the value of the option is calculated under the calibrated risk-neutral measure as follows:

$$\tilde{V}_{t_0} = \sum_{i=1}^{M} p_i^* \exp\left(-r(t_i^* - t_0)\right) CF_{i,t_i^*}, \tag{4.20}$$

where $\tilde{V}_{t_0}$ represents the estimated price of the American-style option.

---

[9] Douady (2001) also proposes to use the exercise value as a further 'basis function' in the regression. Then the matrix $\mathbf{A}$ consists of $(K+3)$ columns, where the elements of the last column are equal to the exercise value.

[10] Note that $\sum_{i=1}^{\widehat{M}} w_{i,i}$ is not necessarily equal to one, since we only sum up over a subset of paths. This problem is resolved by simple normalization.

Note that our choice of $w_{i,i} = p_i^*$ corresponds, in an econometric sense, to the case of heteroscedastic disturbances, where $w_{i,i} = \frac{1}{\sigma_i^2}$ and σ_i^2 represents the variance of disturbances. Therefore, we have $p_i^* = \frac{1}{\sigma_i^2}$ which means that a higher probability corresponds to a lower residual variance of that state.

Obviously, for $w_{i,i} = p_i^* \equiv \frac{1}{M}$ the method coincides exactly with the Longstaff and Schwartz (2001) approach. To avoid numerical overflow errors and to get as precise results as possible, the payoffs CF_{i,t_k} and the stock prices $X_{i,t_k}^{(m)}$ should be normalized by the strike price K as recommended by Longstaff and Schwartz (2001).

In the end, this extended approach still does not allow us to use American-style options to calibrate the model by Avellaneda et al. (2001), since it is not possible to calculate the discounted cash flows of these types of options at time t_0. To calculate the cash flows the optimal probabilities are needed, but they cannot be obtained before the model is calibrated.

4.3.3 Comparison of the Extensions

The major difference between the two extensions proposed above is that the WTA (i.e. the extended Andersen (1999) approach) parameterizes the optimal exercise frontier, whereas in the WLSA (i.e. the extended Longstaff and Schwartz (2001) approach) the continuation value of the option still alive is approximated by regression. Furthermore, in the WTA, the exercise decision at time t_k only depends on the current exercise value, which is compared to the threshold θ_k. In contrast, in the WLSA, the current value of the state variables at t_k on path i influences both the value of continuation and the exercise value.

Let us take a closer look to the exercise decision. In the WTA, the optimal exercise decision is integrated in the optimization routine, whereas, in the WLSA, the estimation of the continuation value function and the exercise decision are separated, i.e. the estimation is independent from the exercise decision at the current time point t_k.

The WTA could generally be used for all types of options. We propose to apply the approach to the pricing of plain-vanilla options, since the simple structure of the products coincides with the simple setup of the approach. Additionally, the method is straightforward to implement since only standard routines are necessary. The main disadvantage of the approach is that the optimization routine to find the optimal path probabilities must be done twice, which can become time-consuming.

Furthermore, the threshold parameter is difficult to interpret in a more sophisticated environment.

These problems do not occur in the WLSA. Here, we are able to model the continuation value function depending on more than one variable. Thus, a scenario analysis can be conducted, e.g. the determination of the optimal exercise decision depending on the asset price and the volatility in a stochastic volatility model. As a result, this approach should be used in a more complex environment. The main drawback of the approach is that numerical problems can emerge, which can influence the accuracy of the results. Note that this is also the case without our extension as reported in Longstaff and Schwartz (2001). The state variables should always be normalized, since some basis functions consist of exponential terms which could lead to numerical underflow/overflow errors. Furthermore, some basis functions are highly correlated, and, therefore, the cross-moment matrices are often nearly singular.[11]

In Sect. 4.5 we apply both approaches to plain-vanilla American put options in a real market environment and compare the results with and without weighting the Monte Carlo paths. Before doing, so we show the importance of weighting the Monte Carlo paths to reproduce the prices of traded options exactly.

4.4 Effect of Weighting Monte Carlo Paths

The main reason why Monte Carlo paths should be weighted is that in real market environments the case of a perfect model fit to market prices will rarely occur, especially for models with low degrees of freedom. For example, in the Black–Scholes model, a unique constant volatility parameter that matches all market prices of traded benchmark options simultaneously does not exist, abstracting from the special case where only one benchmark option is traded. Besides this, most calibration procedures rely on the existence of explicit pricing formulas for the prices of benchmark instruments. However, closed-form solutions for prices are not always available or easy-to-compute. In this case, fitting the model to market prices implies searching the parameter space via direct simulation, which is computationally expensive and time-consuming.

When closed-form expressions exist, the model parameters can often be simply estimated via least-squares methods. If the model para-

---

[11] We use the Numerical Recipes routine SVDFIT recommended by Press et al. (1992) for linear regression problems, which is more stable than using normal equations.

meters can be specified such that all benchmark instruments are correctly priced, then there is no need to weight the Monte Carlo paths, i.e. to change the weights from the uniform case to some new p_i^*. In this section, we will show the effect of weighting Monte Carlo paths by assessing experimentally how much re-weighting is necessary in the ideal case of a perfectly specified model.

We generate prices by the Black–Scholes formula for different strikes and maturities. We consider European options on a stock with a spot price of 100 that pays no dividends. The interest rate is assumed to be 5%. Taking a 'maximum horizon' for the model of 270 days, we use European call options with times to maturity of 30, 60, 90, 180, and 270 days as our benchmark instruments. Strike prices are chosen from 90 to 110 with step size 2.5. We assume that the prices are given by the Black–Scholes formula with a volatility of 20%. Furthermore, we add 'zero-strike' calls for each maturity date to ensure that the model is calibrated to forward prices, and, hence, there is no net bias in the forward prices.[12] As our prior to simulate the sample paths we take a geometric Brownian motion with drift 5% and volatility 20%.

The test consists of 1,000 replications of the calibration algorithm described in Sect. 4.2.2. Therefore, we use in each replication different seed values to generate the sample paths. All Monte Carlo runs were made with 10,000 paths (Type II), consisting of 5,000 paths and their antithetics. We use equal weights $w_j = 10^{-4}$ for the least-squares approximation.[13] For each replication we calculate the value of the entropy measure $D(p|u)$ given in (4.3). Since $D(p|u)$ takes values in the interval $[0, \log(M)]$, we normalize $D(p|u)$ by $\log(M)$ to get values in the interval $[0, 1]$. Thereby, a value of 0 corresponds to uniformly distributed paths, whereas a value of 1 shows that the probability of 1 is assigned to only one path. Figure 4.1 displays the histogram of all normalized values of $D(p|u)$. We find that all values are relatively small, with a median value of 0.000226 and a maximum value of 0.000505. As expected, this indicates that only little weighting is needed. The fact that we need any weighting at all in this setting is due to finite sample errors. Figure 4.2 depicts a typical histogram of the normalized paths probabilities (i.e. normalized by $\frac{1}{M}$).

The distribution of the calibrated probabilities is unimodal and strongly peaked at its mean. This confirms the intuition that the weight-

---

[12] This guarantees that the mean of the distribution of the asset price at the maturity date is fitted exactly.

[13] Therefore, the discrepancy between our 'market' prices and the calibrated prices, which is typically of order $\frac{1}{\sqrt{w_j}}$, is less than one cent.

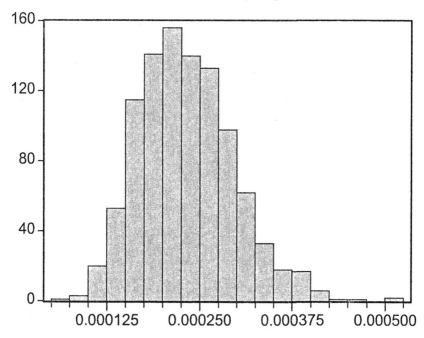

Fig. 4.1. Histogram of the normalized values of $D(p|u)$. The distribution is unimodal with median 0.000226.

ing of Monte Carlo paths is only needed to correct small finite sample errors, if the parameters of the chosen model can be specified exactly. To show this, we repeat the experiment three times. First, we use only 5,000 paths and calibrate to the benchmark prices (Type I), then we use 10,000 paths (Type III), and 100,000 paths (Type IV) without calibration.[14] The results are given in Table 4.1.

It can easily be seen that with weighting the sample paths the relative pricing error is much less than 0.1% for all maturities and strike prices. Furthermore, the number of paths has no significant influence on the relative pricing error. Without calibration using 10,000 paths, the relative pricing errors are much higher. If one compares these results with the relative pricing errors using 100,000 paths, this error is three times smaller than for 10,000 paths.[15]

This confirms our conjecture that path weighting is not needed, if the model parameters can be specified in such a way that the parameters

---

[14] Each time one half of the paths consists of their antithetics.

[15] Note that this coincides with the expected theoretical convergence rate $\frac{1}{\sqrt{M}}$ of Monte Carlo simulations. Since we have used ten times as many paths, the error is reduced by $\frac{1}{\sqrt{10}} \approx \frac{1}{3}$.

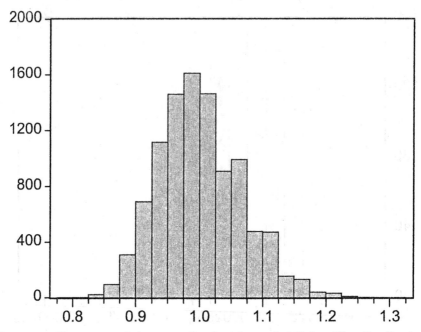

Fig. 4.2. Histogram of the normalized paths probabilities. The distribution is unimodal and strongly peaked at 1.0. Note that there are some outliers. The normalized value of $D(p|u)$ of this repetition was 0.000226.

price all benchmark instruments correctly. In this case, the errors that occur are only due to finite sample errors and usually lie within the bid-ask-spread for a sufficient large number of sample paths. Hence, for correctly specified models a trade-off exists between a 'quick and dirty' price and a more accurate one, which requires more computation time.

In practice, such a correct specification is normally not feasible, since there are more traded benchmark instruments in the market than there are model parameters. Therefore, in real market environments, the parameters can only be estimated via least-squares methods. These estimates should be used to generate the prior, since this is already the best guess for the given model. Afterwards, the approach by Avellaneda et al. (2001) should be used to re-calibrate the prior-distribution, to ensure pricing by simulation in accordance with all benchmark instruments. In Sect. 4.5 we apply the Weighted Monte Carlo approach to market data. There, the relative $D(p|u)$ is 0.0312 for the Black–Scholes model, which is over 130 times higher than the median value for the exactly specified model, and over 60 times higher than the maximum value of the relative $D(p|u)$ for an exactly specified model. This sup-

Table 4.1. Range for Relative Pricing Error

The entries in the table are the ranges for the relative pricing error (in percent) in a single Monte Carlo run. Type I uses 5,000 paths and weighting, Type II 10,000 paths and weighting, Type III 10,000 paths without weighting, and Type IV 100,000 paths without weighting.

Type	Strike	Time to Maturity		
		30 Days	60 - 90 Days	180 - 270 Days
I		[-0.025,0.012]	[-0.032,0.005]	[-0.065,0.018]
II	ITM	[-0.018,0.008]	[-0.029,0.002]	[-0.061,0.022]
III	[90.0,95.0]	[-0.972,0.976]	[-1.687,1.464]	[-2.040,2.334]
IV		[-0.265,0.281]	[-0.481,0.464]	[-0.753,0.775]
I		[-0.026,0.016]	[-0.036,0.010]	[-0.073,0.014]
II	ATM	[-0.023,0.012]	[-0.033,0.006]	[-0.067,0.020]
III	[97.5,102.5]	[-5.258,5.030]	[-5.615,4.638]	[-3.965,4.268]
IV		[-1.749,1.538]	[-1.390,1.410]	[-1.338,1.314]
I		[-0.029,0.018]	[-0.037,0.010]	[-0.071,0.010]
II	OTM	[-0.025,0.012]	[-0.034,0.04]	[-0.068,0.020]
III	[105.0,110.0]	[-18.169,22.470]	[-11.985,13.330]	[-6.833,6.242]
IV		[-5.529,5.903]	[-3.721,3.048]	[-2.262,2.169]

ports the importance of weighting Monte Carlo paths to reproduce the market prices of actively traded benchmark instruments correctly.

4.5 Example

4.5.1 Data

Next, we use our extensions to price plain-vanilla American put options in a real market environment and compare the results with the original approaches by Andersen (1999) and Longstaff and Schwartz (2001). For this purpose, we take all bid and ask quotes for the April, May, June, September, and December DAX index option quoted at EUREX on March 30th, 2004 at 12:00 P.M. CET to construct a set of benchmark instruments. The corresponding times to maturity of these options are 20, 55, 83, 174, and 265 days, respectively. As a proxy for the risk-free rate, we use the 3-month EURIBOR, which was 1.958% p.a. on March 30th. Since the DAX is a performance index with dividend reinvestment, we do not need to take care of the payout rate. The level of the DAX index is taken from XETRA, which provides the best bid and offer prices. The midpoint of the bid and ask quotes was 3864.815 at the time of sampling.

In a next step, general arbitrage violations have to be eliminated from the data, since we want to estimate risk-neutral probability distributions. Therefore, we apply the same method as Jackwerth and Rubinstein (1996) to sort out arbitrage violations in the data. Hence, two options are sorted out so that the final sample consists of 103 call options[16] – 19, 22, 25, 18, and 19 for the different maturity dates, respectively. In addition to these call prices, we add 'zero-strike' calls with prices corresponding to the spot price of the DAX value for the different maturity dates. This is a prerequisite to avoid a net bias in the underlying asset price process. Thus, we have a set of 108 benchmark instruments to calibrate the models.

To measure the performance of the different approaches, we need market quotes for plain-vanilla American puts written on the DAX. However, on EUREX all DAX options are European. In order to get a comparable sample, we take all bid and ask quotes for bank-issued American DAX puts with time to maturity less than or equal to 265 days contemporaneously quoted at EUWAX[17]. The sample consists of 220 plain-vanilla American put options with time to maturity between 9 and 265 days and strike prices ranging from 2000 to 4800. Beside the issuer of each warrant, all other option characteristics are identical.

4.5.2 Model Specification

To derive a 'prior' for the Monte Carlo simulation, we use the models by Black and Scholes (1973) and Heston (1993), and thereafter we recalibrate the 'prior' to the benchmark instruments.

Black and Scholes (1973) (BS) assume that the stock price follows a geometric Brownian motion with constant instantaneous volatility σ, i.e.

---

[16] Puts are translated into calls using European put-call parity.

[17] Bank-issued instruments are non-standardized, and individual issuers are free to choose any option characteristics. The European Warrant Exchange (EUWAX) is a trading segment of the Stuttgart Stock Exchange listing many types of warrants and certificates and is the world's largest derivative exchange in terms of listed bank-issued instruments. Each quote from a market maker represents a minimum trading volume of EUR 3,000 or 10,000 securities. In addition, the market maker is obliged to quote tradable bid and offer prices on a continuous basis during trading hours. Therefore, we conclude that the quality of the quotes in terms of represented liquidity is quite high, which is verified in a recent study by Bartram and Fehle (2004). This study also contains a detailed description of the differences in the market microstructure and market design of the EUREX and the EUWAX. The complete guidelines for the EUWAX, as well as further information are available at www.euwax.com.

$$dS_t = rS_tdt + \sigma S_tdW_t \tag{4.21}$$

where dW_t is the increment of a standard Wiener process. Note that the dynamics of the stock are given under the risk-neutralized probability measure. The only unknown parameter is the instantaneous volatility σ, which has to be estimated. As proxy for σ, we use the implied volatility under which the sum of squared pricing errors is minimized and calculated as 0.22805.

The Heston (1993) model incorporates stochastic variance in addition to the stochastic process of the underlying. The dynamics of the underlying stock and its instantaneous variance under the risk-neutralized measure are given in the model by

$$dS_t = rS_tdt + \sigma_t S_tdW_t^{(1)}, \tag{4.22}$$

$$d\sigma_t^2 = \kappa(\theta - \sigma_t^2)dt + \eta\sigma_tdW_t^{(2)}, \tag{4.23}$$

where the increments of the Wiener processes $dW_t^{(1)}$ and $dW_t^{(2)}$ are correlated with constant correlation coefficient ρ. The parameters κ, θ, and η represent the speed of adjustment, the long-run mean, and the volatility of the instantaneous variance σ_t^2. Heston (1993) has shown that by using Fourier inversion techniques a closed-form pricing formula to the model can be derived. Therefore, it is convenient to estimate the unknown parameters κ, θ, η, ρ, and the instantaneous variance $\sigma_{t_0}^2$ for the given benchmark instruments via least-squares methods. The parameters for our data are $\kappa = 3.25695$, $\theta = 0.079038$, $\eta = 0.71281$, $\rho = -0.84036$, and $\sigma_{t_0}^2 = 0.04678$, i.e. the local volatility at time t_0 is 0.21628.

We use the estimated parameters for the BS model and the Heston model to simulate 50,000 paths of (4.21), as well as of (4.22) and (4.23), consisting of 25,000 paths and their antithetics. Thus, we choose one time step per day for the discretization of the processes, i.e. $\Delta t = \frac{1}{365}$. The continuous exercise features for the American options are thus approximated with one exercise decision per day. Afterwards, we calibrate the simulated prices to the market prices as described in Sect. 4.2.2 to get the 'calibrated' risk-neutral measure. We use $w_j = \frac{10^{-4}}{C_j}$ for the least-squares approximation in (4.5). We obtain a normalized value $D(p|u)$ for the BS model of 0.03122 and of 0.0101 for the Heston (1993) model. It is no surprise that the Heston model yields a better prior for the market data, since there are more free parameters for calibration.

Afterwards, we price the American put options traded at EUWAX using the original threshold approach by Andersen (1999), the original least-squares approach by Longstaff and Schwartz (2001), and our extensions presented above in Sect. 4.3.

For the threshold approaches, the early exercise boundary is determined by generating 25,000 paths and their antithetics, using the same parameters for the respective model as described above. Furthermore, we apply the extended version proposed by Douady (2001), i.e. we use (4.14) and (4.15) in the optimization problem (4.11). Numerical tests indicate that $\alpha = 1,000$ is a good choice for the fuzziness parameter. Additionally, we use normalized values for $IV_{i,k}$ and $H_{i,k}$, i.e. we divide by the current asset value S_{t_0}.

For the least-squares approaches, we have to choose the basis functions to approximate the continuation value. For the BS model, we follow Longstaff and Schwartz (2001) and use a constant term and the first three Laguerre polynomials $n = 0, 1, 2$ given in (4.16). For the Heston (1993) model, we choose as basic functions in the regression a constant term, the first two Laguerre polynomials evaluated at the DAX index level, the first two Laguerre polynomials evaluated at the volatility level, and the cross products of the first three Laguerre polynomials evaluated at the DAX index level and at the volatility level. Thus, we have a total of eight basis functions. The volatility level is included in the regression, since the exercise decision depends on both the index level and the volatility level.[18]

4.5.3 Empirical Results

To assess the performance of the different approaches three criteria are used. These are defined as follows:

- The root mean squared percentage error (RMSPE) is the square root of the average squared percentage deviations of the simulated values from the reported midpoint option prices.
- The mean outside percentage error (MOPE) is the average percentage error outside the bid-ask spread. The outside error is defined as the difference between the bid (ask) price and the simulated value if the simulated value is below (above) the reported option bid (ask) price. It is set equal to zero otherwise. Therefore, a positive (negative) value of MOPE means that the simulated values are too high (low) on average.

---

[18] The volatility is similar to the stock price in the case of valuing American-Asian options as described in Longstaff and Schwartz (2001), Sect. 4. Here the immediate exercise value of the option depends only on the average stock price, but the stock price has an influence on the future average of the stock price, and, therefore, an influence on the exercise decision.

- The mean absolute outside percentage error (MAOPE) is the average absolute percentage valuation error outside the bid-ask spread. This measure illustrates how accurately each approach fits within the quoted bid and ask prices for each option category.

Table 4.2 contains the results using the Black–Scholes model as the prior model for the Monte Carlo simulation. The original approaches by Andersen (1999) and Longstaff and Schwartz (2001), referred to as 'Standard', perform similarly without weighting the Monte Carlo paths. However, the former is faster than the latter, especially for long-term deep-in-the-money options, while the least-squares approach requires a lot of computation time. Using our proposed extensions, referred to as 'Weighted', we can reduce the RMSPE by about one third for the least-squares approach and by one fourth for the threshold approach. Similar results are obtained for the MAOPE. The MOPE is negative in all cases, which verifies the hypothesis that both approaches give a lower bound for the correct price.

Table 4.2. Pricing Error Statistics using the Black–Scholes Model

RMSPE is the root mean squared percentage valuation error averaged over all 220 plain-vanilla American puts traded at EUWAX on March 30th. MOPE is the average percentage valuation error outside the bid-ask spread. MAOPE is the average absolute percentage valuation error outside the bid-ask spread.

| | Least-Squares Approach Longstaff and Schwartz (2001) | | Threshold Approach Andersen (1999) | |
	Standard	Weighted	Standard	Weighted
RMSPE	0.28315	0.20544	0.28623	0.22646
MOPE	-0.11480	-0.07204	-0.11759	-0.11230
MAOPE	0.13265	0.09739	0.13560	0.11377

We now take a look at the results for the Heston model in Table 4.3. Roughly speaking, the RMSPE for the 'Standard' approaches is half of what we found for the Black–Scholes model. In addition, accuracy is improved using our extended approaches. The other results are comparable to the findings for the Black–Scholes model.

Summarizing the results, we recommend the use of a more sophisticated model as a 'benchmark' model to generate the Monte Carlo paths, which can reproduce a volatility smile. First, using a well-specified model to run the Monte Carlo simulation reduces the RMSPE significantly. Second, the weighting algorithm proposed by Avellaneda et al. (2001) is not able to remedy all model misspecifications. Our approaches

Table 4.3. Pricing Error Statistics using the Heston Model

RMSPE is the root mean squared percentage valuation error averaged over all 220 plain-vanilla American puts traded at EUWAX on March 30th. MOPE is the average percentage valuation error outside the bid-ask spread. MAOPE is the average absolute percentage valuation error outside the bid-ask spread.

| | Least-Squares Approach Longstaff and Schwartz (2001) | | Threshold Approach Andersen (1999) | |
	Standard	Weighted	Standard	Weighted
RMSPE	0.15830	0.12285	0.16318	0.13045
MOPE	-0.06473	-0.04563	-0.06912	-0.06055
MAOPE	0.06574	0.05000	0.06960	0.06179

to price plain-vanilla American put options reduces the RMSPE in both models as compared with the original approaches. In addition, both least-squares approaches, the original approach by Longstaff and Schwartz (2001) and our extended version, perform better than the corresponding threshold approach by Andersen (1999) and our extension. However, one has to keep in mind that the least-squares approaches are more time-consuming than the threshold approaches. Therefore, a trade-off exists between a 'quick and dirty' price, which can be calculated using the threshold approaches, and more precise results generated by the least-squares approaches.

Note that the problem of parameter specification and the choice of a 'good' model can be mitigated by using our proposed extensions to simulate American option prices, since we combine the non-parametric approach by Avellaneda et al. (2001) with the approach by Andersen (1999) and Longstaff and Schwartz (2001) to approximate the value of American-style derivatives. Therefore, we conclude that re-calibration improves model robustness.

4.6 Conclusions

In this chapter, we propose two new flexible approaches to price American-style derivatives in accordance with given market prices. To this aim, we combine the weighted Monte Carlo approach for calibrating Monte Carlo simulations proposed by Avellaneda et al. (2001) with the Least-Squares Monte Carlo approach suggested in Longstaff and Schwartz (2001) and with the threshold approach proposed by Andersen (1999). Both approaches are intuitive, easy-to-apply and computationally efficient. We illustrate the original methods and our extensions using the example for valuing standard American put options with market data from EUREX and EUWAX. We find that the pricing performance depends on two factors: the chosen prior model to generate the Monte Carlo paths and the application of our extensions. In addition, the least-squares approaches perform better than the corresponding threshold approaches, but they are also more time-consuming. Therefore, the user has the choice between a 'quick and dirty' price and more precise prices using the least-squares approaches. Our extensions mitigate model misspecification, which supports the benefit of the two-step procedure to regard market information.

Further research should examine how these methods perform when more complex derivatives (e.g. path-dependent American options or other exotic options) are priced. In particular, no clear prediction exists of the structure and form of the polynomials and which kind of stochastic processes should be used. Moreover, the question of which model is the 'best' still remains open. All these issues certainly warrant further research.

References

Achieser, N.I. and I.M. Glasmann (1981): *Theorie der linearen Operatoren im Hilbert-Raum, 8. Aufl.*, Deutsch, Frankfurt/Main.

Aït-Sahalia, Y. and A.W. Lo (1998): Nonparametric Estimation of State-Price Densities Implicit in Financial Asset Prices, *The Journal of Finance*,
53(2):499–547.

Andersen, L. (1999): A Simple Approach to the Pricing of Bermudan Swaptions in the Multi-Factor LIBOR Market Model, *The Journal of Computational Finance*, 3(2):5–32.

Avellaneda, M., R. Buff, C. Friedman, N. Grandchamp, L. Kruk, and J. Newman (2001): Weighted Monte Carlo: A New Technique for Calibrating Asset-Pricing Models, *International Journal of Theoretical and Applied Finance*, 4(1):91–119.

Barraquand, J. and D. Martineau (1995): Numerical Valuation of High-Dimensional Multivariate American Securities, *Journal of Financial and Quantitative Analysis*, 30(3):383–405.

Bartram, S.M. and F.R. Fehle (2004): *Alternative Market Structures for Derivatives*, EFA 2003 Annual Conference Paper No. 297.

Black, F. and M. Scholes (1973): The Valuation of Options and Corporate Liabilities, *Journal of Political Economy*, 81(3):637–654.

Boyle, P., M. Broadie, and P. Glasserman (1997): Monte Carlo Methods for Security Pricing, *Journal of Economic Dynamics and Control*, 21:1267–1321.

Broadie, M. and P. Glasserman (1997): Pricing American-Style Securities Using Simulation, *Journal of Economic and Dynamic Control*, 21:1323–1352.

Brown, G. and K.B. Toft (1999): Constructing Binomial Trees from Multiple Implied Probability Distributions, *The Journal of Derivatives*, 7(2):83–100.

Buchen, P. and M. Kelly (1996): The Maximum Entropy Distribution of an Asset Inferred from Option Prices, *The Journal of Financial and Quantitative Analysis*, 31(1):143–159.

Carriere, J. (1996): Valuation of Early-Exercise Price of Options using Simulations and Nonparametric Regression, *Insurance: Mathematics and Economics*, 19(1):19–30.

Derman, E. and I. Kani (1994): Riding on a Smile, *RISK*, 7(2):32–39.

Douady, R. (2001): Bermudan Option Pricing with Monte-Carlo Methods, in Avellaneda, M., editor, *Quantitative Analysis in Financial Markets – Vol. III* World Scientific Publishing Co., 314–328.

Duffie, D. (2001): *Dynamic Asset Pricing Theory, 3. ed.*, Princeton University Press, Princeton.

Dumas, B., J. Fleming, and R.E. Whaley (1998): Implied Volatility Functions: Empirical Tests, *The Journal of Finance*, 53(6):2059–2106.

Dupire, B. (1994): Pricing with a Smile, *RISK*, 7(1):18–20.

Greene, W.H. (1999): *Econometric Analysis, 4. ed.*, Prentice Hall, New Jersey.

Gulko, L. (1999): The Entropy Theory of Stock Option Pricing, *International Journal of Theoretical and Applied Finance*, 2(3):331–355.

Gulko, L. (2002): The Entropy Theory of Bond Option Pricing, *International Journal of Theoretical and Applied Finance*, 5(4):355–383.

Heston, S. (1993): A Closed-Form Solution for Options with Stochastic Volatility with Applications to Bond and Currency Options, *The Review of Financial Studies*, 6(2):327–343.

Heuser, H. (1992): *Funktionalanalysis, 3. Aufl.*, Teubner, Stuttgart.

Ibáñez, A. and F. Zapatero (2004): Monte Carlo Valuation of American Options Through Computation of the Optimal Exercise Frontier, *The Journal of Financial and Quantitative Analysis*, 39(2):253–275.

Jackwerth, J.C. (1997): Generalized Binomial Trees, *The Journal of Derivatives*, 5(2):7–17.

Jackwerth, J.C. and M. Rubinstein (1996): Recovering Probability Distributions from Option Prices, *The Journal of Finance*, 51(5):1611–1631.

Johnston, J. and J. DiNardo (1997): *Econometric Methods, 4. ed.*, McGraw-Hill, New York.

Longstaff, F. and E. Schwartz (2001): Valuing American Options by Simulation: A Simple Least-Squares Approach, *The Review of Financial Studies*, 14(1):113–147.

Musiela, M. and M. Rutkowski (1997): *Martingale Methods in Financial Modelling*, Springer, Berlin – Heidelberg – New York.

Niederreiter, H. (1992): *Random Number Generation and Quasi-Monte Carlo Methods*, SIAM, Philadelphia.

Press, W.H., S.A. Teukolsky, W.T. Vetterling, and B.P. Flannery (1992): *Numerical Recipes in C: The Art of Scientific Computing, 2. ed.*, Cambridge University Press, Cambridge.

Rubinstein, M. (1994): Implied Binomial Trees, *The Journal of Finance*, 49(3):771–818.

Tilley, J.A. (1993): Valuing American Options in a Path-Simulation Model, *Transactions of Society of Actuaries*, 45:499–520.

Wilmott, P. (1999): *Derivatives: The Theory and Practice of Financial Engineering*, John Wiley & Sons, Chichester.

5

Synopsis

5.1 Thematische Einordnung

Geschäfte mit Derivaten haben Hochkonjunktur, da Derivate den gezielten Handel von Risiken erlauben. So setzen Hedgefonds Derivate sehr häufig zur gezielten Übernahme von Risiken ein. Durch die große Anzahl von weltweit zugelassenen Hedgefonds, die immer komplexere Produkte nachfragen, wird dieser Boom verstärkt. In Deutschland wurden mit dem Investmentmodernisierungsgesetz im Jahre 2003 auch die Voraussetzungen für die Zulassung von Hedgefonds in Deutschland geschaffen sowie die Genehmigung erteilt, Derivate zur Absicherung von Sondervermögen im Bereich der Publikumsfonds einzusetzen. Im Gegensatz zu der gezielten Übernahme von Risiken werden derivative Finanzinstrumente aber auch als Mittel zur Risikoreduktion in vielen Wirtschaftszweigen eingesetzt. Eine Erhebung der International Swaps and Derivatives Association (ISDA) aus dem Jahre 2003 zeigt, dass über 90% der 500 größten Unternehmen weltweit Derivate im Risikomanagement einsetzen. Davon betroffen ist auch die bilanzielle Behandlung von Finanzinstrumenten. Hinsichtlich der Bewertung für Derivate gilt in der angloamerikanischen Bilanzpraxis (IAS 39, US-GAAP 133), die von den großen Unternehmen angewendet wird, der Grundsatz der Marktbewertung ('mark-to-market' bzw. 'Fair-Value'). Darüber hinaus ist auch für die Berichterstattung der Fondsgesellschaften ein Ansatz der Derivate zum Marktwert erforderlich. Problematisch wird der Ansatz zum Marktwert, wenn es sich um außerbörslich gehandelte Derivate, sog. OTC-Derivate, handelt. Für diese wird im Allgemeinen kein Marktpreis notiert, sodass entsprechende Bewertungsverfahren notwendig sind.

Die bestehenden Ansätze zur Bewertung von Optionen können in zwei Klassen unterteilt werden. Die erste Klasse bilden parametrische Optionsbewertungsmodelle. Diese traditionellen Ansätze zur Bewertung von Derivaten gehen von einem datengenerierenden Prozess aus, d.h., es wird ein stochastischer Prozess für den Preis des Underlyings angenommen. Die populärste Wahl für diesen datengenerierenden Prozess ist die geometrische Brownsche Bewegung, erstmalig verwendet in der Optionspreistheorie von Black und Scholes (1973). Empirische Studien (z.B. Rubinstein (1994), Jackwerth und Rubinstein (1996), Dumas et al. (1998) oder Aït-Sahalia und Lo (1998)) zeigen jedoch, dass die implizite Volatilität, die durch Invertierung der Black–Scholes Formel bestimmt wird, in Abhängigkeit vom Basispreis (sog. 'volatility smiles' oder 'volatility skews') und in Abhängigkeit von der (Rest-)Laufzeit ('term structure of volatility') der Optionen variiert, während das BS Modell von einer konstanten Volatilität ausgeht. Damit wird die Gültigkeit des Modells in Frage gestellt. Um den Zusammenhang zwischen impliziter Volatilität und Basispreis und/oder Laufzeit abbilden zu können, wurden erweiterte parametrische Optionsbewertungsmodelle vorgeschlagen. Dazu zählen Modelle mit stochastischer Volatilität (z.B. Hull und White (1987), Heston (1993)), Schöbel and Zhu (1999)), mit stochastischen Zinsen (z.B. Merton (1973), Amin und Jarrow (1992)) oder mit stochastischen Sprüngen (z.B. Merton (1976), Bates (1991)), sowie Kombinationen aus den verschiedenen stochastischen Prozessen (z.B. Bates (1996), Scott (1997), Bakshi und Chen (1997)). Nach der Auswahl eines geeigneten Modells werden die Preisprozesse direkt unter dem äquivalenten Martingalmaß angegeben. Anschließend werden die Parameter der Prozesse bestimmt. Dies geschieht durch geeignete Schätzverfahren aus gegebenen Marktpreisen. Nachdem die Parameter der stochastischen Prozesse spezifiziert worden sind, wird der Preis eines Derivates als Funktion der Parameter sowie der aktuellen Preise der Underlyings bestimmt.

In den meisten Fällen können diese Modelle die beobachteten Marktpreise von gehandelten Derivaten jedoch nicht exakt reproduzieren (vgl. Das und Sundaram (1999), Belledin und Schlag (1999)). Daher sollten sie nur mit äußerster Sorgfalt in der Praxis angewandt werden, insbesondere bei der Bewertung und Absicherung von exotischen Optionen. Aufgrund der Tatsache, dass zur Absicherung von exotischen Optionen sehr häufig europäische Standardoptionen verwendet werden, kann die Hedgingperformance durch eine konsistente Bewertung wesentlich verbessert werden.

Daher wurden in der Literatur neue Ansätze vorgeschlagen, die die Betrachtung umkehren. Diese marktkonformen Ansätze bzw. impliziten Verfahren zur Bewertung von Optionen bilden die zweite Klasse der Optionsbewertungsmodelle. Dieses Bewertungsprinzip wird auch als konsistente Bewertung oder als 'Fair-Value'-Ansatz bezeichnet. Die Preise von Derivaten werden dabei in Abhängigkeit von am Markt beobachteten Preisen von bereits gehandelten und vom Markt bewerteten Derivaten bestimmt. Dabei werden die Marktpreise der gehandelten Optionen als richtig bewertet angenommen, um daraus Informationen über den Prozess des Underlyings zu gewinnen.[1] Die Implementierung dieser Ansätze erfolgt überwiegend in einem zeit- und zustandsdiskreten Modellrahmen, da numerische Methoden zu deren Umsetzung angewendet werden. Die bekanntesten Verfahren sind implizite Binomial-/Trinomialbäume, implizite finite Differenzen-Verfahren sowie gewichtete Monte Carlo Simulationen.

In dieser Dissertation liegt der Fokus auf der Entwicklung neuer Ansätze zur marktkonformen Bewertung von Derivaten. Dabei liegt ein besonderes Augenmerk auf der Bewertung von illiquiden Optionen sowie auf neu einzuführenden Optionen. Ausgehend von bestehenden Verfahren werden neue Ansätze vorgeschlagen und deren Anwendung anhand von Beispielen demonstriert.

5.2 Struktur und Inhalt der Arbeit

Im ersten Kapitel wird ein neuer Ansatz zur Konstruktion von arbitragefreien impliziten Binomialbäumen vorgeschlagen, der auf dem Ansatz von Brown und Toft (1999) basiert. Das Verfahren ermöglicht die Konstruktion eines arbitragefreien, impliziten Binomialbaumes, der sowohl mit der 'term structure' der impliziten Volatilitäten als auch mit dem 'volatility smile' bzw. dem 'volatility skew' konsistent ist. Der Hauptvorteil dieser Methode liegt darin, dass die impliziten Wahrscheinlichkeitsverteilungen (IRNPDs) für spätere Fälligkeitstermine endogen aus den IRNPDs von vorherigen Fälligkeitsterminen innerhalb eines Optimierungsproblems bestimmt werden. Dies stellt eine wesentliche Erweiterung des Ansatzes von Brown und Toft (1999) dar. Diese verwenden ein dreistufiges Verfahren, um einen arbitragefreien, impliziten Binomialbaum zu konstruieren. Jedoch werden bei dieser Methode nicht alle vorhandenen Informationen optimal verarbeitet, da die IRNPDs

---

[1] Die Kalibrierung der traditionellen Ansätze an gegebene Marktpreise wird in dieser Arbeit nicht als implizites Verfahren betrachtet.

für jeden Fälligkeitstermin separat geschätzt werden. Anschließend werden – durch Lösung eines weiteren Optimierungsproblems – bedingte implizite Verteilungen zwischen zwei aufeinanderfolgenden Fälligkeitsterminen bestimmt. In einigen Fällen hat dieses Optimierungsproblem jedoch keine Lösung, da die Nebenbedingungen nicht erfüllt werden können.

Im Gegensatz dazu liefert das Optimierungsproblem in dem neu vorgestellten Ansatz immer eine Lösung. Darüber hinaus kann das Optimierungsproblem in ein quadratisches Optimierungsproblem mit linearen Nebenbedingungen transformiert werden, welches mit Standardroutinen gelöst werden kann.

Ein weiterer Vorteil dieses neuen Verfahrens liegt in der Arbitragefreiheit des impliziten Binomialbaumes, die per Konstruktion sichergestellt ist. Dadurch ist garantiert, dass keine negativen Übergangswahrscheinlichkeiten wie in den Ansätzen von Derman und Kani (1994) und Barle und Cakici (1998) auftreten. Ferner werden keine Interpolationsverfahren benötigt, sondern nur tatsächlich beobachtete Preise als Inputparameter verwendet. Dieses neue Verfahren erlaubt – nach unserem Kenntnisstand – erstmalig implizite Binomialbäume von der Wurzel beginnend zu konstruieren, in denen alle Informationen optimal verarbeitet werden und die per Konstruktion arbitragefrei sind. Dieses war bisher nur mit dem Ansatz von Jackwerth (1997) möglich, der aber per Rückwärtsinduktion konstruiert wird und die häufige Lösung eines nicht-linearen Optimierungsproblems erfordert.

Durch Einführung einer Volatilitätsfunktion kann die Methode ebenfalls verwendet werden, um arbitragefreie, implizite Multinomialbäume zu konstruieren. Multinomialbäume können verwendet werden, um mehr als eine Zustandsvariable zu erfassen wie z.B. die Preise von zwei unterschiedlichen Underlyings, um damit Korrelationsprodukte zu bewerten. Darüber hinaus stellen Multinomialbäume eine Verfeinerung von Binomialbäumen dar und liefern daher genauere Resultate bei derselben Anzahl von Zeitschritten.

Eine Unterkategorie der Multinomialbäume sind Trinomialbäume. Typisch für implizite Trinomialbäume ist, dass der komplette Zustandsraum für den Baum im voraus spezifiziert wird und nur die Übergangswahrscheinlichkeiten bestimmt werden müssen. Zur Konstruktion dieser Bäume nach den Verfahren von Dupire (1994) und Derman et al. (1996) ist jedoch ebenfalls eine Interpolation der Marktpreise notwendig. Daraus folgt wiederum, dass auch in diesen impliziten Trinomialbäumen negative Übergangswahrscheinlichkeiten vorkommen können. Dies führt ebenfalls zu nummerischen Ungenauigkeiten und Instabilitäten wie bei den impliziten Binomialbäumen von Derman und Kani (1994) und Barle und Cakici (1998).

Im zweiten Kapitel dieser Arbeit wird der neu vorgestellte Ansatz in einem empirischen Performancetest mit anderen Verfahren verglichen. Dazu werden insgesamt sechs unterschiedliche Optionsbewertungsmodelle miteinander verglichen: das *'klassische'* Black–Scholes Modell, das stochastische Volatilitätsmodell von Heston (1993), zwei deterministische Volatilitätsmodelle, nämlich die Spezifikation einer Volatilitätsfunktion und die Anwendung einer einfachen Interpolationsmethode, die häufig in der Praxis vorkommt, das Umgewichten von Monte Carlo Pfaden nach dem Ansatz von Avellaneda et al. (2001) zur marktkonformen Optionsbewertung mittels Monte Carlo Simulation, sowie der neu vorgeschlagene Ansatz zur Konstruktion impliziter Binomialbäume. Im Gegensatz zu früheren Studien z.B. von Bakshi et al. (1997), Dumas et al. (1998) oder Jackwerth und Rubinstein (2001) steht dabei die Bewertungsperformance der Modelle im Vordergrund und nicht, wie in den erwähnten Studien, die Zeitkonsistenz bzw. die Hedgingperformance der Modelle. Zu diesem Zweck wird jedes Modell an gegebene Marktpreise kalibriert, die an der EUREX im Zeitraum von Januar 2004 bis Juni 2004 beobachtet wurden. Dabei wird die Kalibrierung für jeden Handelstag neu vorgenommen.

Ziel einer marktkonformen Bewertung ist es, illiquide oder neu einzuführende Optionen zu bewerten. Um dieses Ziel abbilden zu können, werden Daten von der EUWAX verwendet, an der sog. *'Covered Warrants'*, von Investmentbanken ausgegebene Optionsscheine, gehandelt werden. Dabei wird angenommen, dass diese Optionen neu zu emittierende Optionen darstellen. Die an die EUREX-Daten kalibrierten Modelle werden dann verwendet, um an der EUWAX gehandelte Optionen zu bewerten. Die jeweiligen Modellpreise werden mit den beobachteten Preisen an der EUWAX verglichen und entsprechend ausgewertet. Als neu zu emittierende Optionen werden dabei amerikanische Kaufoptionen und Knock-Out Optionen verwendet.

Der Performancetest zeigt, dass die Bewertung mittels impliziter Binomialbäume sowohl für amerikanische Kaufoptionen als auch für Knock-Out Optionen die besten Ergebnisse liefert. Die Umgewichtung von Monte Carlo Pfaden liefert ebenfalls gute Ergebnisse. Auffallend ist, dass eine Umgewichtung der Pfade, die auf Grundlage des Black–Scholes Modells erzeugt werden, schlechte Resultate für die Bewertung von Up-and-Out-Verkaufsoptionen liefern. Wie erwartet zeigt das Black–Scholes die schlechteste Performance, da es auch die geringste Anzahl an Parametern zur Kalibrierung besitzt. Die Ergebnisse der deterministischen Volatilitätsmodelle sowie des Modells von Heston sind gemischt und liegen zwischen den Extremen.

Insgesamt wurde bei diesem Test festgestellt, dass die beobachteten Marktpreise an der EUWAX im Durchschnitt höher waren als die berechneten Modellpreise. Dies ist gleichzusetzen mit einer Unterbewertung und wurde für alle verwendeten Modelle beobachtet. Dies ist zurückzuführen auf die Unterschiede in der Marktmikrostruktur der beiden Optionsmärkte EUREX und EUWAX. Während an der EUREX überwiegend institutionelle Investoren zur Absicherung ihrer Positionen als Marktteilnehmer auftreten, so handeln an der EUWAX überwiegend Privatinvestoren mit Spekulationsmotiven. Damit einhergehend ist das Transaktionsvolumen entsprechend geringer.

Die Resultate des empirischen Tests zeigen, dass implizite Binomialbäume sehr gut zur marktkonformen Bewertung von Optionen geeignet sind. Die Anwendung impliziter Binomialbäume ist jedoch beschränkt auf eindimensionale Problemstellungen. Ferner können implizite Binomialbäume auch nur begrenzt zur Bewertung pfadabhängiger Optionen eingesetzt werden. Zur Bewertung mehrdimensionaler Probleme, d.h. Probleme mit mehr als einer stochastischen Zustandsvariable wie z.B. Bewertung von Multi-Asset Optionen oder Basket-Optionen, sowie zur Bewertung pfadabhängiger Optionen, werden i.d.R. Monte Carlo Simulation verwendet, da diese einfach zu implementieren sind und dabei eine hohe Flexibilität aufweisen. Um entsprechende Simulationen durchführen zu können, wird zuerst ein parametrisches Optionsbewertungsmodell ausgewählt und anschließend werden die Parameter der stochastischen Prozesse festgelegt. Damit können die beobachteten Marktpreise jedoch i.d.R. nicht perfekt repliziert werden. Daher haben Avellaneda et al. (2001) einen neuen Ansatz zur Kalibrierung von Monte Carlo Simulationen vorgeschlagen. Um die beobachteten Marktpreise exakt abzubilden, wird jedem Pfad eine Pfadwahrscheinlichkeit zugeordnet, sodass sich der Preis einer Option als gewichteter, diskontierter Mittelwert der Auszahlungen auf jedem Pfad ergibt, wobei die Pfadwahrscheinlichkeiten als Gewichte bei der Berechnung dieses Wertes verwendet werden. Die Pfadwahrscheinlichkeiten sind dabei so zu bestimmen, dass die gewichteten, diskontierten Mittelwerte der Zahlungen den beobachteten Martkpreisen entsprechen und Abweichungen von der Gleichverteilung minimal sind. Somit ist es möglich, eine marktkonforme Bewertung mittels Monte Carlo Simulation durchzuführen.

Bei der Verwendung von Monte Carlo Simulationen besteht jedoch noch ein weiteres Problem, das lange Zeit als unlösbar galt, nämlich die Bewertung von amerikanischen Optionen. Dies ist bedingt durch das Simulationsprinzip, das, vom aktuellen Wert beginnend, mögliche Zustände in der Zukunft simuliert, wohingegen amerikanische Optionen mittels Rückwärtsinduktion bewertet werden. Aufgrund der großen Bedeutung amerikanischer Optionen wurden in den letzten Jahren eine Vielzahl von Ansätzen zur Bewertung amerikanischer Optionen mittels Monte Carlo Simulation vorgeschlagen, z.B. Tilley (1993), Andersen (1999), Longstaff und Schwartz (2001) oder Ibánez und Zapatero (2004).

Im dritten Kapitel werden daher erstmalig zwei Ansätze zur marktkonformen Bewertung von amerikanischen Optionen mittels Monte Carlo Simulation vorgeschlagen. Zu diesem Zweck wird der Ansatz von Avellaneda et al. (2001) auf der einen Seite mit dem Ansatz von Andersen (1999) kombiniert, sowie auf der anderen Seite mit dem 'Least-Squares' Monte Carlo Ansatz von Longstaff und Schwartz (2001). Beide Ansätze sind intuitiv und somit einfach anzuwenden. Die Anwendung dieser beiden neuen Ansätze sowie der Originalvorschläge von Andersen (1999) und Longstaff und Schwartz (2001) werden anhand der Bewertung von amerikanischen Verkaufsoptionen mit Marktdaten von der EUREX und EUWAX für einen Handelstag in einem kleinen Beispiel veranschaulicht. Dabei zeigt sich, dass die gewichteten Ansätze wie erwartet bessere Ergebnisse liefern als die Verwendung der ungewichteten Originalansätze. Die Performance hängt dabei von zwei Faktoren ab: zum einen von dem zugrunde gelegten parametrischen Optionsbewertungsmodell sowie von der Anwendung der neu eingeführten Ansätze. Ferner stellt man fest, dass die 'Least-Squares' Methode etwas bessere Resultate liefert als die Methode von Andersen (1999), jedoch dafür auch mehr Rechenzeit in Anspruch nimmt. Folglich hat der Anwender die Wahl zwischen einem 'quick and dirty' Preis oder einem genaueren Preis, dessen Berechnung jedoch mehr Zeit in Anspruch nimmt.

5.3 Ausblick

In dieser Arbeit wurden verschiedene Ansätze zur marktkonformen Bewertung von Optionen vorgeschlagen. Dabei wurde im zweiten Kapitel in einem ausführlichen Performancetest festgestellt, dass Optionen, die an der EUWAX gehandelt werden, anscheinend eine Prämie enthalten. In diesem Zusammenhang erscheint eine weitergehende Analyse dieser Bewertungsunterschiede angebracht. Insbesondere ist dabei die Frage von Interesse, ob die Preisunterschiede nur von den Options-Charakteristika abhängen oder auch abhängig vom jeweiligen Emittenten sind.

Die marktkonforme Bewertung von amerikanischen Optionen mittels Monte Carlo Simulation wurde nur anhand eines einfaches Beispiels demonstriert. Zukünftige Studien sollten daher einen ausführlichen Performancetest dieser Methoden über einen längeren Zeitraum und mit komplexeren Produkten untersuchen. Dabei ist von besonderem Interesse, welche stochastischen Prozesse zur Erzeugung der Monte Carlo Pfade verwendet werden sollten.

References

Aït-Sahalia, Y. and A.W. Lo (1998): Nonparametric Estimation of State-Price Densities Implicit in Financial Asset Prices, *The Journal of Finance*, 53(2):499–547.

Amin, K. and R. Jarrow (1992): Pricing Options on Risky Assets in a Stochastic Interest Rate Economy, *Mathematical Finance*, 2(4):217–237.

Andersen, L. (1999): A Simple Approach to the Pricing of Bermudan Swaptions in the Multi-Factor LIBOR Market Model, *The Journal of Computational Finance*, 3(2):5–32.

Avellaneda, M., R. Buff, C. Friedman, N. Grandchamp, L. Kruk, and J. Newman (2001): Weighted Monte Carlo: A New Technique for Calibrating Asset-Pricing Models, *International Journal of Theoretical and Applied Finance*, 4(1):91–119.

Bakshi, G., C. Cao, and Z. Chen (1997): Empirical Performance of Alternative Option Pricing Models, *The Journal of Finance*, 52(5):2003–2049.

Bakshi, G. and Z. Chen (1997): An Alternative Valuation Model for Contingent Claims, *Journal of Financial Economics*, 44(1):123–165.

Barle, S. and N. Cakici (1998): How to Grow a Smiling Tree, *Journal of Financial Engineering*, 7(2):127–146.

Bates, D. (1991): The Crash of '87: Was It Expected? The Evidence from Options Markets, *The Journal of Finance*, 46(3):1009–1044.

Bates, D. (1996): Jumps and Stochastic Volatility: Exchange Rate Processes Implicit in Deutsche Mark Options, *The Review of Financial Studies*, 9(1):69–108.

Belledin, M. and C. Schlag (1999): An Empirical Comparison of Alternative Stochastic Volatility Models, *Working Paper Series Finance & Accounting, Fachbereich Wirtschaftswissenschaften*, Goethe-Universität, Frankfurt am Main

Black, F. and M. Scholes (1973): The Valuation of Options and Corporate Liabilities, *Journal of Political Economy*, 81(3):637–654.

Brown, G. and K.B. Toft (1999): Constructing Binomial Trees from Multiple Implied Probability Distributions, *The Journal of Derivatives*, 7(2):83–100.

Das, S.R. and R.K. Sundaram (1999): Of Smiles and Smirks: A Term Structure Perspective, *The Journal of Financial and Quantitative Analysis*, 34(2):211–239.

Derman, E. and I. Kani (1994): Riding on a Smile, *RISK*, 7(2):32–39.

Derman, E., I. Kani, and N. Chriss (1996): Implied Trinomial Trees of the Volatility Smile, *The Journal of Derivatives*, 3(4):7–22.

Dumas, B., J. Fleming, and R.E. Whaley (1998): Implied Volatility Functions: Empirical Tests, *The Journal of Finance*, 53(6):2059–2106.

Dupire, B. (1994): Pricing with a Smile, *RISK*, 7(1):18–20.

Heston, S. (1993): A Closed-Form Solution for Options with Stochastic Volatility with Applications to Bond and Currency Options, *The Review of Financial Studies*, 6(2):327–343.

Hull, J.C. and A. White (1987): The Pricing of Options on Assets with Stochastic Volatilities, *The Journal of Finance*, 42(2):281–300.

Ibánez, A. and F. Zapatero (2004): Monte Carlo Valuation of American Options Through Computation of the Optimal Exercise Frontier, *The Journal of Financial and Quantitative Analysis*, 39(2):253–275.

Jackwerth, J.C. (1997): Generalized Binomial Trees, *The Journal of Derivatives*, 5(2):7–17.

Jackwerth, J.C. and M. Rubinstein (1996): Recovering Probability Distributions from Option Prices, *The Journal of Finance*, 51(5):1611–1631.

Jackwerth, J.C. and M. Rubinstein (2001): Recovering Stochastic Processes from Option Prices, *Working Paper*, London Business School.

Longstaff, F. and E. Schwartz (2001): Valuing American Options by Simulation: A Simple Least-Squares Approach, *The Review of Financial Studies*, 14(1):113–147.

Merton, R.C. (1973): Theory of Rational Option Pricing, *Bell Journal of Economics and Management*, 4:141–183.

Merton, R.C. (1976): Option Pricing when Underlying Stock Returns are Discontinuous, *Journal of Financial Economics*, 3:125–144.

Rubinstein, M. (1994): Implied Binomial Trees, *The Journal of Finance*, 49(3):771–818.

Schöbel, R. and J. Zhu (1999): Stochastic Volatility With an Ornstein-Uhlenbeck Process: An Extension, *European Finance Review*, 3(1):23–46.

Scott, L.O. (1997): Pricing Stock Options in a Jump-Diffusion Model with Stochastic Volatility and Interest Rates: Applications of Fourier Inversion Methods, *Mathematical Finance*, 7(4):413–426.

Tilley, J.A. (1993): Valuing American Options in a Path-Simulation Model, *Transactions of Society of Actuaries*, 45:499–520.

Lecture Notes in Economics and Mathematical Systems

For information about Vols. 1–475
please contact your bookseller or Springer-Verlag

Printing and Binding: Strauss GmbH, Mörlenbach